# FLORA OF TROPICAL EAST AFRICA

## OCHNACEAE

BERNARD VERDCOURT

Trees, shrubs or geophytic shrublets, less often annual or perennial herbs. Leaves alternate, simple (pinnate and opposite in *Rhytidanthera*, a small S American genus) often with numerous lateral veins and densely reticulate tertiary venation, entire to serrate or setulose; stipules entire to deeply divided, deciduous or persistent. Flowers hermaphrodite, almost always regular, solitary or in fascicles or raceme-like, paniculate or cymose inflorescences; pedicels often articulated. Sepals (3–4–)5(–6–10), free, usually quincuncially imbricate, persistent or deciduous. Petals (4–)5(–6–12), free, contorted in bud, often clawed, deciduous. Stamens few to many, with persistent filaments; anthers linear, basifixed, dehiscing lengthwise or by apical pores; staminodes present outside stamens in a few genera (*Sauvagesia* and allies). Ovary superior, sessile, syncarpous, entire to lobed with style and 2–5 parietal placentas each with 1–many ovules, or with (3–)5(–6–15) lobes each with 1 ovule and style gynobasic; styles as many as placentas or ovary-lobes but completely united or free only at the apex with one globose or lobed stigma or separate stigmas. Fruit a collection of 3–12 one-seeded drupelets borne on a fleshy enlarged receptacle, a nut surrounded by unequal enlarged sepals or a septicidal 2–5-valved, 1–many-seeded capsule. Seeds with or without endosperm.

A family of 26 genera and about 360 species, widespread in the tropics and subtropics of both Old and New Worlds. Corner, Seeds Dicotyledons 1: 208 (1976) was convinced that *Sauvagesia* and its allies belong to a separate family not closely related to Ochnaceae but on account of fruit and seed characters close to Violaceae as suggested by Benth. & Hook. G.P. 1: 114–121 (1862). *Lophira* was formerly placed in the Dipterocarpaceae. M.C.E. Amaral (E.J. 113: 105–196 (1991)) has given an account of the phylogenetic systematics of the family. She confirms that the two genera mentioned above belong in the Ochnaceae but her paper does not help in the decision of the division of the *Ochneae* e.g. the *Gomphia/Ouratea* problem.

1. Annual or perennial herbs; seeds with endosperm; fruit a 3-valved septicidal capsule . . . . . . . . . . . . . . 5. **Sauvagesia** (p. 55)
   Trees, shrubs or sometimes short woody-based pyrophytes, never true herbs; seeds without endosperm; fruit not capsular . . . . . . . . . . . . . . . . . . . . . . . . . . . . . . 2
2. Fruit a nut surrounded by very unequal enlarged wing-like sepals; large-leaved savanna tree of U 1 . 4. **Lophira** (p. 53)
   Fruit of 1–many mostly black drupelets borne on an enlarged usually red receptacle with persistent ± equal usually red sepals . . . . . . . . . . . . . . . . . . . . . . . . . . . . 3
3. Stamens 10 with anthers much longer than filaments, dehiscing by apical pores; petals yellow to orange . . . . . . . . . . . . . . . . . . . . . . . . . . . . 3. **Gomphia** (p. 43)
   Stamens 13 or more with anthers longer or shorter than filaments dehiscing by longitudinal slits or apical pores or if 8–10 then anthers equalling filaments and dehiscing longitudinally; petals yellow to orange, white or pink . . . . . . . . . . . . . . . . . . . . . . . . . . 4

1

© The Board of Trustees of the Royal Botanic Gardens, Kew, 2005

4. Stipules entire, bifid or fringed, not striate, usually
   soon deciduous; petals pale yellow to orange (very
   rarely white); drupelets without internal
   projection of endocarp . . . . . . . . . . . . . . . . . . . .       1. **Ochna**
   Stipules laciniate or deeply divided into linear
   segments, markedly longitudinally striate,
   persistent on young shoots; petals white to pink;
   drupelets with internal projection of endocarp;
   yellow pigment usually evident beneath bark . . . .       2. **Brackenridgea** (p. 39)

## 1. OCHNA

L., Sp. Pl.: 513 (1753) & Gen. Pl. ed. 5: 229 (1754)

Trees, shrubs or pyrophytic shrublets, usually glabrous. Young leaves often reddish. Leaves petiolate, usually with serrate or ciliate margins, less often entire; stipules entire to deeply bifid or fimbriate, usually not striate nor persistent. Inflorescences paniculate, umbellate or racemiform, sometimes reduced to a solitary flower, terminal or terminating short axillary shoots; usually scented; inflorescence buds usually very noticeable, fusiform, ovoid or subglobose with numerous distichous imbricate bracts leaving a series of annular scars which are usually very obvious; pedicels articulated at or above the base; torus usually thickening and becoming red in fruit. Sepals (3–)5, quincuncially imbricate, becoming red and coriaceous in fruit. Petals 5(–12), absent in one species, mostly yellow, usually unguiculate. Stamens (14–)20–numerous in 2 or more whorls, free; filaments slender, longer or shorter than the anthers, persistent; anthers yellow, dehiscing by longitudinal slits or terminal pores, usually soon deciduous. Carpels (3–)5–15, free, 1-ovulate; styles slender, gynobasic, united or sometimes free at apex with somewhat enlarged stigmas. Fruit with 1–several free black drupelets with fleshy mesocarp. Seeds straight, ± curved or reniform, without endosperm; embryo straight or curved.

An Old World genus of about 80 species, nearly all in Africa and Madagascar, one in Mauritius and four in Asia extending from India and Sri Lanka to Indochina, Hainan, Nicobar Is. and Malay Peninsula. A record from Timor is erroneous. The various types of embryos are figured in F.C.B. Ochnaceae: 3, fig. 1 (1967).

I have followed Robson's clear division of the genus into three sections:
- Sect. *Renicarpus*. Carpels and drupelets reniform attached by middle      (sp. 1)
- Sect. *Ochna* (*Diporidium* (Bartl. & Wendl.) Engl.). Anthers with pores      (spp. 2–21)
- Sect. *Schizanthera* Engl. Anthers with slits                               (spp. 22–36)

Spp. 37–47 are those I have been unable to match and think may be new taxa but anthers are not available.

It cannot be pretended that naming *Ochna* species is easy. Without young flowers with anthers it is not possible to use the key effectively although by using both leads results can be obtained. Characters are not always constant. In couplet 23 for example some of the species may have shoots and/or pedicels glabrous rather than papillose-puberulous. To make allowances for all exceptions would make the key too unwieldy. A list is given of the species occurring in each Flora area, which may help in identification.

1. Drupelets (and carpels) distinctly reniform, attached
   by the middle; inflorescence characteristically
   narrowly paniculate, 5–12 cm long (**U** 2, 4; **K** 3; **T** 1)       1. *O. membranacea* (p. 7)
   Drupelets (where known) ± ellipsoid or subglobose,
   attached at base or close to base (in *O.* sp. 39
   distinctly above the base); inflorescences various,
   flowers often few . . . . . . . . . . . . . . . . . . . . . . . . . . . . . . . . . . . . . . . . . . . . . . . . . . . 2

© The Board of Trustees of the Royal Botanic Gardens, Kew, 2005

2. Anthers dehiscing by apical pores; styles usually
    shortly free at apex . . . . . . . . . . . . . . . . . . . . . . . . . . . . . . . . . . . . . . . . . . . . . 3
   Anthers dehiscing by longitudinal slits; styles usually
    completely united . . . . . . . . . . . . . . . . . . . . . . . . . . . . . . . . . . . . . . . . . . . . . . 22
3. Petals absent; leaves oblanceolate to elliptic, 6.5–14
    (–17) × 1.7–4(–5) cm, midrib yellow-brown in life,
    drying yellowish beneath (**K** 7; **T** 3, 8) . . . . . . . . . . .  2. *O. apetala* (p. 8)
   Petals present . . . . . . . . . . . . . . . . . . . . . . . . . . . . . . . . . . . . . . . . . . . . . . . . . . 4
4. Leaf margins with few basal fine setae or with fine setae
    all round; usually coastal (but sometimes to 700 m) . . . . . . . . . . . . . . . . . . . . 5
   Leaf margins toothed or serrulate, the teeth triangular
    or curved etc. (and perhaps with terminal seta) but
    not finely setiform or ± entire but lacking basal setae . . . . . . . . . . . . . . . . 7
5. Leaves thin, narrowly elliptic, oblong-elliptic or
    oblanceolate, 2.3–7.5(–12) × 0.7–2.2(–3.5) cm; petals
    smaller, 9–15 × 7–9 mm; carpels 5 (**K** 7; **T** 3, 6, 8) . . .  3. *O. holtzii* (p. 9)
   Leaves more coriaceous; petals larger; carpels 8–12 . . . . . . . . . . . . . . . . . . . . . . . 6
6. Leaves 2–15 × 1–5 cm; petals 12–22 × 8–15 mm;
    marginal setae 1–4 mm long, often restricted to
    margins near base (**K** 7; **T** 3, 6; **Z**; **P**) . . . . . . . . . . . .  4. *O. thomasiana* (p. 9)
   Leaves larger 5–21 × 2.5–7 cm; petals 13–25 × (6–)
    10–18 mm; marginal setae up to 2 mm long,
    numerous around margin (often black-tipped when
    dry) (**K** 7; **T** 3, 6) . . . . . . . . . . . . . . . . . . . . . . . . . . . . .  5. *O. kirki*
                                                  subsp. *multisetosa* (p. 10)
7. Bud scales persistent on the young stems, ± imbricate,
    lanceolate, 2–5 mm long, 1–1.5 mm wide; flowers
    solitary (**U** 2, 4) . . . . . . . . . . . . . . . . . . . . . . . . . . . . .  13. *O. bracteosa* (p. 17)
   Bud scales deciduous or only odd ones ± persistent . . . . . . . . . . . . . . . . . . . . 8
8. Anthers long, 4–8 mm long . . . . . . . . . . . . . . . . . . . . . . . . . . . . . . . . . . . . . . . 9
   Anthers short, 1–3 or if up to 5.5 mm then leaf
    margins entire or with only small areas of serration
    and leaf-bases rounded to cordate . . . . . . . . . . . . . . . . . . . . . . . . . . . . . . 15
9. Petals large, at least some usually over 2 cm long . . . . . . . . . . . . . . . . . . . . . 10
   Petals smaller, 9–15(–20) × 4–9 mm . . . . . . . . . . . . . . . . . . . . . . . . . . . . . . . . 12
10. Lateral veins with area of lamina around them often
     conspicuously impressed, ± bullate when dry; shrub
     or small tree (widespread in **K** 1, 3–6; also in **U** 4
     and **T** 2) . . . . . . . . . . . . . . . . . . . . . . . . . . . . . . . . . . . . .  8. *O. insculpta* (p. 14)
    Lateral veins usually not as above (use geography if
     in doubt) . . . . . . . . . . . . . . . . . . . . . . . . . . . . . . . . . . . . . . . . . . . . . . . . . . 11
11. Shrub or small tree or rhizomatous shrublet
     0.05–5(–9) m tall; flowers numerous in branched
     panicles terminating lateral shoots, the 4–8 mm
     long basal pedicel joints forming characteristic tufts
     when flowers have fallen (**K** 1, 7; **T** 3, 6, 8; **Z**) . . . . .  6. *O. mossambicensis* (p. 11)
    Suffrutex or shrub (or small tree in **K** 7) 0.1–3 m tall,
     flowers (2)3–9(–14)) in racemiform inflorescences
     with short rhachis not falling to leave tufts of
     pedicel joints but fruiting peduncles ± thick (**K** 7;
     **T** 1, 3–8) . . . . . . . . . . . . . . . . . . . . . . . . . . . . . . . . . .  7. *O. macrocalyx* (p. 12)
12. Flowers on leafless branches on spur shoots so closely
     placed that they form a tight 30–40-flowered cluster
     at apices of branches; shrub 3.6 m; leaves not seen
     completely adult but narrowly oblong-lanceolate
     about 9 × 2.3 cm (**T** 6, Kiserawe Forest Reserve) . . .  9. *O. polyarthra* (p. 16)
    Flowers not as above . . . . . . . . . . . . . . . . . . . . . . . . . . . . . . . . . . . . . . . . . . 13

© The Board of Trustees of the Royal Botanic Gardens, Kew, 2005

13. Leaves obovate, 13–15 × 6–8 cm, rounded at apex;
    filaments 3–4 mm long, anthers 5 mm long (**T** 8,
    Lindi) ........................................ 10. *O. citrina* (p. 16)
    Leaves lanceolate or oblanceolate, acute or acuminate
    to a subobtuse apex; filaments 1–4 mm long,
    anthers 5.5–7 mm long ........................................ 14
14. Leaves lanceolate, 5.5–8(–10.5) × 1.2–2.2 cm, acute
    (**T** 8, Lindi) ........................................ 11. *O. braunii* (p. 16)
    Leaves oblanceolate, 9–16 × 3–4(–5) cm, acuminate
    to a subobtuse apex (**T** 8, Lindi) ................ 12. *O. schliebenii* (p. 17)
15. Anthers 3–5.5 mm long; filaments 4–7 mm long;
    leaves elliptic-oblong, 3.3–10 × 1.4–4.5 cm, rounded,
    subcordate or cordate at the base, completely
    entire or with short setae (**T** 6, 8) .............. 5. *O. kirkii*
                                                           subsp. *kirkii* (p. 10)
    Without these characters combined; anthers 1–2.5 mm
    long ........................................ 16
16. Petals 26 × 15 mm; fruiting sepals up to 3 cm long,
    tightly imbricate around fruit, only sometimes
    ultimately spreading; pedicels jointed at base (**T** 6, 8) 21. *O. rovumensis* (p. 22)
    Petals much smaller (not seen for sp. 17) ........................... 17
17. Flowers apparently not precocious ................................. 18
    Flowers definitely precocious ................................. 20
18. Leaves lanceolate or oblong-elliptic, 3.5–6.5 × 1–2.4 cm,
    acutely acuminate at the apex (**T** 8) ............. 20. *O. pseudoprocera* (p. 22)
    Leaves acute, subacute or rounded at the apex ...................... 19
19. Shrub of fringing forest; petals 9 × 6 mm; fruiting sepals
    spreading (**U** 2, 4) ........................... 18. *O. hackarsii* (p. 20)
    Shrub of open woodland, petals 5–7 mm long;
    fruiting sepals spreading (**U** 1) ................ 19. *O. leucophloeos* (p. 21)
20.\*Intricately branched shrub or small tree usually with
    small elliptic or round leaves 1.7–4.8 × 0.9–2.8 cm
    (in East Africa), rounded or obtuse at apex; flowers
    solitary; petals 7–11 mm long; fruiting sepals usually
    remaining imbricated around the fruit ........... 14. *O. inermis* (p. 18)
    More openly branched shrub or small tree with larger
    leaves; flowers in 1–8-flowered inflorescences
    sometimes aggregated into panicles 5 cm diameter
    or more elongated; fruiting sepals spreading ...................... 21
21. Coastal plant with more oblong leaves and fascicles
    4–8-flowered (**K** 7, Lamu below 5 m) ............. 17. *O.* sp. (p. 20)
    Widespread plant up to 2100 m, usually with broadly
    elliptic leaves with very convex sides; fascicles
    1–3(–4) flowered, often aggregated to form panicles
    (**U** 1; **K** 1, 4, 6, 7; **T** 1–8) ..................... 16. *O. ovata* (p. 19)
22. Evergreen forest tree up to 27 m tall; flowers 5–20 in
    racemiform inflorescences with rhachis up to 2 cm
    long; jointed part of pedicels ± 3 mm long, forming
    characteristic persistent clusters (any *Ochna* of
    timber size is likely to be this species but it is
    sometimes shrubby and these forms can be difficult
    to identify) ........................................ 22. *O. holstii* (p. 23)
    Smaller trees, shrubs or rhizomatous shrublets ...................... 23

---

\* 15. *O. monantha* Gilg will key near here (ex descript.) and may be a synonym of *O. inermis* or
a form of *O. ovata*. See p. 19.

© The Board of Trustees of the Royal Botanic Gardens, Kew, 2005

23. Young branches and pedicels papillose-puberulous
    (often needs × 10 lens or better to see) . . . . . . . . . . . . . . . . . . . . . . . . . . . . 24
    Young branches and pedicels glabrous (see note at
    head of key) . . . . . . . . . . . . . . . . . . . . . . . . . . . . . . . . . . . . . . . . . . . . . 27
24.*Leaves thin, acute or acutely acuminate, 2–6.7 × 1–2
    cm; branches slender with numerous short leafy
    shoots; upland evergreen forest 1650–2400 m (T 6, 7)   23. *O. oxyphylla* (p. 24)
    Leaves narrowly rounded at the apex or subacute;
    more woodland or wooded grassland plants but also
    in forest remnants . . . . . . . . . . . . . . . . . . . . . . . . . . . . . . . . . . . . . . . 25
25. Inflorescence rhachis usually well-developed, the
    inflorescence (5–)6–10(–14)-flowered; leaves some-
    times drying dark bluish green, often bronze-tinged
    when young, (2.5–)3.5–6(–12) × 1–3(–5) cm;
    fruiting sepals spreading (T 6–8) . . . . . . . . . . . . . .   25. *O. polyneura* (p. 25)
    Inflorescence usually not well-developed or if so then
    fruiting sepals remaining tightly imbricate around
    fruit, only eventually sometimes ± spreading . . . . . . . . . . . . . . . . . . . . . 26
26. Fruiting sepals spreading, 5–6 mm wide; flowers not
    precocious (1800–2100 m, T 7) . . . . . . . . . . . . . . . .   24. *O. stolzii* (p. 25)
    Fruiting sepals remaining tightly imbricate around
    fruit for most of fruiting period, 8–11 mm wide;
    leaves slightly glaucous often drying dark bluish
    green (1050–2100 m, T 1, 4, 5, 7) . . . . . . . . . . . . . .   26. *O. puberula* (p. 26)
27. Leaves ± glaucous, very broadly oblong-elliptic, up to
    14.5 × 8.5 cm, very broadly rounded or even
    emarginate at apex; several racemiform 7–16-
    flowered inflorescences often forming globose
    clusters 8–15 cm wide; sepals 16–25 mm long, convex;
    branchlets thick (T 4, 7) . . . . . . . . . . . . . . . . . . . . . .   32. *O. gambleoides* (p. 31)
    Without these characters . . . . . . . . . . . . . . . . . . . . . . . . . . . . . . . . . . . . 28
28. Small pyrophytic subshrubs 0.1–1.3(–2) m tall . . . . . . . . . . . . . . . . . . . . . . 29
    Shrubs or trees, rarely only 1.5 m tall (some glabrous
    specimens of *O. polyneura* may key here – so if no
    satisfactory answer, return to couplet 23 and try
    other alternative) . . . . . . . . . . . . . . . . . . . . . . . . . . . . . . . . . . . . . . . . . 32
29. Leaves obovate to obovate-oblanceolate, subcoriaceous
    and mostly ± glaucous, (3–)4–12(–14) × 1.2–4.5
    (–5.2) cm, usually entire or remotely serrulate; shoots
    often caespitose greyish white with often peeling
    papery epidermis; flowers usually precocious (T 1,
    4, 6–8) . . . . . . . . . . . . . . . . . . . . . . . . . . . . . . . . . . . . .   28. *O. leptoclada* (p. 28)
    Leaves less coriaceous, not glaucous, distinctly serrulate;
    shoots not white with peeling bark . . . . . . . . . . . . . . . . . . . . . . . . . . . . 30
30. Anthers 1–1.5 mm long; petals 7–10 mm long;
    flowers ± precocious; shrub or shrublet 0.1–1(–2) m
    tall (T 7) . . . . . . . . . . . . . . . . . . . . . . . . . . . . . . . . . . .   35. *O. confusa* (p. 35)
    Anthers 1.8–3 mm long . . . . . . . . . . . . . . . . . . . . . . . . . . . . . . . . . . . . . 31
31. Pyrophytic subshrub 25–45 cm tall, typically forming
    cushions from a many-headed woody rootstock;
    sometimes flowering at ground level . . . . . . . . . . . .   30. *O. katangensis* (p. 29)
    Subshrub to 60 cm with more simply branched stems   31. *O. richardsiae* (p. 31)

* Some specimens with larger leaves will key here which have been doubtfully referred to
shrubby forms of *O. holstii* (see note at end of sp. 22).

© The Board of Trustees of the Royal Botanic Gardens, Kew, 2005

32.*Leaves slightly distinctly acute or acuminate or
     narrowed or acuminate to a very narrowly rounded
     apex . . . . . . . . . . . . . . . . . . . . . . . . . . . . . . . . . . . . . . . . . . . . . . . . . . . . . . . . . . 33
     Leaves narrowly to broadly rounded at apex . . . . . . . . . . . . . . . . . . . . . . . . . 34
33. Inflorescences (5–)7–10-flowered with distinct rhachis
     9–20 mm long (**K** 2?; **T** 1, 4, 7?) . . . . . . . . . . . . . . . 27. *O. afzelioides* (p. 28)
     Inflorescences subumbellate with 2–6(–8) flowers,
     rhachis 0–4 mm long (**U** 1, 3, 4; **T** 1, 7) . . . . . . . . . . 34. *O. afzelii* (p. 34)
34. Leaves drying dark bluish green, 5.5–12 × 2.3–5; carpels
     6–8 (**T** 5, 7) . . . . . . . . . . . . . . . . . . . . . . . . . . . . . . . . 29. *O. cyanophylla* (p. 29)
     Leaves pale, 5.5–13.5(–17.5) × 1.3–6.5 cm; flowers
     precocious; carpels 5; bark thick, corky, corrugated or
     fissured (**U** 1, 3; **T** 1, 4–8, common and widespread) . . 33. *O. schweinfurthiana* (p. 33)

Synopsis of species for which anthers not known:

1. Small tree from **T** 6, Mafia I. with oblanceolate to oblanceolate-
    elliptic leaves 3.5–13.5 × 1–4.2 cm; flowers and fruits unknown
    (in evergreen forest) . . . . . . . . . . . . . . . . . . . . . . . . . . . . . . . 37. *O.* sp. (p. 35)
    Not from Mafia I. . . . . . . . . . . . . . . . . . . . . . . . . . . . . . . . . . . . . . . . . . . . . . . . . . . 2
2. Inflorescences ± 25-flowered (**T** 6, Rufiji District) . . . . . . . . . . . 45. *O.* sp. (p. 38)
    Inflorescences 1–9-flowered . . . . . . . . . . . . . . . . . . . . . . . . . . . . . . . . . . . . . . . . 3
3. Flowers solitary . . . . . . . . . . . . . . . . . . . . . . . . . . . . . . . . . . . . . . . . . . . . . . . . . . . . 4
    Flowers (1–)2–5 on short spur shoots . . . . . . . . . . . . . . . . . . . . . . . . . . . . . . . 6
4. Flowers solitary at apices of slender side-shoots 2–3 cm long;
    pedicels 2–3 cm long, jointed ± 1 mm from base; tree to 20 m
    with leaves 7 × 2–3 cm, drying dark with black petiole (**T** 6,
    Uluguru Mts) . . . . . . . . . . . . . . . . . . . . . . . . . . . . . . . . . . . 47. *O.* sp. (p. 39)
    Flowers solitary on short leafy shoots; pedicels 1–2.3 cm long,
    jointed 1–4 mm from base; small trees or shrubs ± 3 m high . . . . . . . . . . . 5
5. Lenticels obscure; stipules minute, not fimbriate; leaves 1.2–4.5 ×
    0.5–2 cm; pedicels jointed ± 1 mm from base (**T** 8, Rondo
    Plateau) . . . . . . . . . . . . . . . . . . . . . . . . . . . . . . . . . . . . . . . 38. *O.* sp. (p. 36)
    Lenticels obvious save on youngest shoots; stipules 2–3 mm long
    with short fimbriae at apex; leaves 2.5–6 × 1.2–3; pedicels
    jointed 2–4 mm from base (**T** 6, Ruaha R.) . . . . . . . . . . . . . . 39. *O.* sp. (p. 36)
6. Forest tree to 15 m; flowers 2–5, inflorescence axis ± 5 mm long;
    enlarged fruiting sepals ± round or broadly ovate, imbricate,
    12 × 11–12 mm (**U** 2, Kasyoha-Kitomi Forest) . . . . . . . . . . . 40. *O.* sp. (p. 37)
    Smaller trees with fewer flowers per inflorescence; fruiting sepals
    more oblong or elliptic, narrower . . . . . . . . . . . . . . . . . . . . . . . . . . . . . . . . . . 7
7. Leaves at apex very narrowly acute or narrowly long-acuminate . . . . . . . . . . 8
    Leaves ± rounded at apex or more bluntly acute . . . . . . . . . . . . . . . . . . . . . . 9
8. Pedicels 1.5 cm long, jointed close to base (save for apical
    flower); inflorescence at least 9–flowered; stipules 11–13.5 mm
    long, chaffy, very deciduous (**T** 7, Udzungwa Mts) . . . . . . . . 46. *O.* sp. (p. 39)
    Pedicels 2.5–3 cm long, jointed ± 4 mm from base; flowers 2–3 on
    very short spur shoots; stipules not known (**T** 6, Mikumi
    National Park) . . . . . . . . . . . . . . . . . . . . . . . . . . . . . . . . . . 41. *O.* sp. (p. 37)
9. Bark thick, brown, that on branchlets corky and fissured; leaves
    narrowly elliptic-oblong, 6.5–11 × 2–3.2 mm (**T** 6, Uzaramo,
    under 100 m) . . . . . . . . . . . . . . . . . . . . . . . . . . . . . . . . . . . 42. *O.* sp. (p. 37)
    Bark not thick and corky . . . . . . . . . . . . . . . . . . . . . . . . . . . . . . . . . . . . . . . . . . 10

* 36. *O.* sp. of which the leaves and fruits are unknown will key to here; it is a small tree with
silvery grey trunk from **T** 5 Dodoma, Itigi (p. 35).

© The Board of Trustees of the Royal Botanic Gardens, Kew, 2005

10. Lenticels same colour as pale chestnut brown young shoots; young branchlets and pedicels glabrous; stipules linear-lanceolate, up to 12 mm long, not fimbriate at apex; leaves 2–7.5 × 1–2.8 cm, rather coarsely crenate-serrate (**T** 7, Iringa, Lake Ngwazi) . . . . . . . . . . . . . . . . . . . . . . . . . . . . . . . . . . . . 43. *O.* sp. (p. 38)

    Lenticels much paler than dark purplish stems; young branchlets and pedicels sometimes very minutely papillate; stipules ± 1 mm long, sometimes fimbriate at apex; leaves 3.5–11 × 1.2–4 cm, more finely crenate-serrate (**T** 1, Mwanza) . . . . . . . . . . . . . . . 44. *O.* sp. (p. 38)

List of species occurring in the Flora areas

**U** 1: 14, 16, 19, 28, 33, 34
**U** 2: 1, 13, 18, 40
**U** 3: 13, 22, 33, 34
**U** 4: 1, 8, 13, 18, 34
**K** 1: 4, 6, 8, 14, 16, 22
**K** 2: 14, 16, 27
**K** 3: 1, 8, 16, 22
**K** 4: 8, 14, 16, 22
**K** 5: 8, 16, 22
**K** 6: 8, 16, 22
**K** 7: 2, 3, 4, 5, 6, 7, 14, 16, 17, 22
**T** 1 : 1, 7, 16, 26, 27, 28, 33, 34, 44
**T** 2: 8, 15, 16, 22
**T** 3: 2, 3, 4, 5, 6, 7, 16, 22
**T** 4: 7, 16, 22, 26, 27, 28, 30, 31, 32, 33, 34
**T** 5: 7, 16, 22, 26, 29, 33, 36
**T** 6: 5, 6, 7, 9, 14, 16, 21, 22, 23, 25, 28, 33, 37, 39, 41, 42, 45, 47
**T** 7: 7, 14, 15, 16, 22, 23, 24, 25, 26, 28, 29, 30, 31, 32, 33, 34, 35, 43, 46
**T** 8: 2, 3, 5, 6, 7, 10, 11, 12, 16, 20, 21, 22, 25, 28, 33, 38
**Z** : 4, 6
**P** : 4, 5

1. **Ochna membranacea** *Oliv.* in F.T.A. 1: 316 (1868); Gilg in E.J. 33: 234 (1903); F.W.T.A. ed. 2, 1: 222 (1954); Bamps, F.C.B. Ochnaceae: 16 (1967). Types: Sierra Leone, Sugar Loaf Mt, *Barter* s.n. (K!, syn.) & Nigeria, Idah, *Barter* 1643 (K!, syn.)

Shrub or small tree 0.9–7(–12) m tall with greyish or blackish-purplish striate branchlets; buds fusiform, 10–15 mm long, 2–3 mm in diameter, the bracts entire, deciduous; slash yellowish. Leaves obovate, oblanceolate or elliptic, 5–20 cm long, 1.5–7 wide, narrowly acuminate at the apex, cuneate to somewhat rounded at the base, curved-serrulate; lateral veins ± 15, gently curving upwards, prominent above and beneath but in depressed grooves above (at least when dry); tertiary venation very finely reticulate, slightly prominent but not very evident; stipules linear, 3–15 mm long, up to 5 mm wide, entire or denticulate-fimbriate at the base, deciduous. Flowers in rather narrow panicles 5–12 cm long, with many short side branches ± 5 mm long, each with 5–6 flowers. Pedicels 5–10 mm long, sometimes lengthening to 16 mm in fruit, jointed 1–3 mm from base. Sepals ovate, obovate-elliptic or almost round, (2–)4–5 mm long, 2–3 mm wide, becoming red and enlarging to 16 mm long, 12 mm wide in fruit. Petals yellow, oblanceolate-oblong or spathulate, 5–7 mm long, 1.5–3 mm wide. Anthers 1.5–2 mm long, equalling the filaments, dehiscing by apical pores. Carpels 5; style 4 mm long; stigma capitate. Drupelets black, compressed-reniform, 10–11 mm long, 7–8 mm wide, 4–5.5 mm thick, attached near the middle. Seeds curved; embryo incumbent, heterocotylous with small cotyledon internal.

© The Board of Trustees of the Royal Botanic Gardens, Kew, 2005

UGANDA. Toro District: Muhangi Forest Reserve, 10 June 1952, *A.M. Smith* 34!; Ankole District: Buwezu forest, 1905, *Dawe* 478!; Mengo District: Mabira Forest, Kyagwe [Chagwe], Feb. 1908, *Ussher* 35! & 36!

KENYA. Trans Nzoia District: Mt Elgon, *T.H.E. Jackson* 417!

TANZANIA. Bukoba District: Minziro Forest, Kere Hill, 6 May 1994, *Congdon* 370!

DISTR. **U** 2, 4; **K** 3; **T** 1; Sierra Leone to Cameroon, Central African Republic, Congo (Kinshasa), Sudan, Angola

HAB. Evergreen forest and its margins; 1050–1250 m (–2250 m, see note)

SYN. *O. gilgiana* Engl. in E.J. 33: 243 (1903); F.P.N.A. 1: 613 (1948). Type: Cameroon, Bipindi, near Mimfia, *Zenker* 2336 (B†, holo.)
  *O. tenuissima* Stapf in K.B. 1906: 78 (1906). Type: Uganda, Mengo District: Entebbe, *E. Brown* 345 (K!, holo.)
  *O.* sp.; Dawe, Report Bot. Miss. Uganda Prot.: 40 (1906)
  *O. smythei* Hutch. & Dalz. in F.W.T.A. ed. 1, 1: 190 (1927) & in K.B. 1928: 216 (1928). Type: Sierra Leone, Heirakohum, *Smythe* 124 (K!, holo.)
  *O.\*elegans* sensu Hutch. & Dalz., F.W.T.A. ed. 1, 1: 190 (1927) & in K.B. 1928: 216 (1928) pro parte, *non Monelasmum elegans* Tiegh.
  *O. barteri* sensu Hutch. & Dalz., F.W.T.A. ed. 1, 1: 190 (1927) pro parte, *non Ochnella barteri* Tiegh.

NOTE. The Kenya specimen is sterile but named by Robson and I think correctly so. The flowers are stated to be white and the altitude given as 1950–2250 m and height 9–12 m. Confirmation is very much needed.

2. **Ochna apetala** *Verdc.* **sp. nov.** in sect. *Ochna* ponenda nulla affinitate arcte obvia sed combinatione ut videtur petalorum nullorum cum inflorescentiis parvis 10–25-floris, antheris 3.5 mm longis, costa folii flavo-brunnea distinguenda. Type: Kenya, Kwale District, W Gongoni Forest, *Luke* 2419 (K!, holo., EA, MO, US, iso.)

Shrub 1.5–3 m or tree to 8 m; stems green and flattened in life, purplish-grey and ridged on drying, rather sparsely lenticellate. Leaves fairly thin to subcoriaceous, oblanceolate to elliptic, 6.5–14(–17) cm long, 1.7–4(–5) cm wide, acuminate to an acute apex, cuneate at the base, rather distantly to rather closely serrulate; midrib yellow-brown in life and drying yellowish beneath; lateral veins ± 25, together with many fainter intermediaries and reticulate tertiary venation ± prominent on both surfaces; apical leaf pair on short lateral shoots often ± sub-opposite; petiole 5–7 mm long; stipules narrowly triangular, 3.5 mm long, deciduous. Flowers 10–25 in small condensed, axillary or pseudoterminal inflorescences 2 cm long which are sometimes closely aggregated; bracts chestnut, oblong, 3 mm long, closely striate; pedicels 0–12 mm long, jointed 1–2 mm from the base; buds at first globose, later oblong. Sepals 4–5, oblong, 7.5–10 mm long, 2–3 mm wide, obtuse, becoming red in fruit, 9–10 mm long, 4 mm wide. Petals absent, not found even in bud. Stamens ± 13; anthers 3.5 mm long dehiscing by apical pores; filaments ± 1 mm long. Carpels 4–5; style 5-branched at apex. Drupelets 3–5, black, subglobose, compressed, 7–9 mm long, 5–6 mm wide, 4 mm thick.

KENYA. Kwale District: Gongoni Forest Reserve, NE side, 2 June 1990, *Robertson & Luke* 6348! & Buda Mafisini Forest Reserve, 23 Feb. 1989, *Luke & Robertson* 1674!; Kilifi District: SW slope of Kaya Jibana, 14 Dec. 1990, *Luke & Robertson* 2639!

TANZANIA. Tanga District: Pande, 19 Aug. 1982, *Hawthorne* 1456!; Lindi District: Chitoa Forest Reserve, 18 June 1995, *Clarke* 68!

DISTR. **K** 7; **T** 3, 8; not known elsewhere

HAB. Lowland rain-forest of *Cynometra, Julbernardia, Diospyros, Parkia, Fernandoa* etc., semi-deciduous forest of *Afzelia, Julbernardia, Craibia, Milicia* etc., also grassland with *Hyphaene* and sometimes riverine; in **T** 8 'dry forest ridge'; 5–200 (240–400 in **T** 8) m

* Given erroneously as *Ouratea* in F.W.T.A. ed. 2.

© The Board of Trustees of the Royal Botanic Gardens, Kew, 2005

NOTE. Luke mentions on his 2419 (Kwale District, W Gongoni Forest, 9 June 1990) that there are no petals, not even in the buds. A bud I dissected certainly had none, though a minute insect larva had eaten several anthers and pollen. I do not think this can account for the missing petals. Confirmation is needed from other localities that this is a constant feature of this species.

3. **Ochna holtzii** *Gilg.* in E.J. 33: 244 (1903); T.T.C.L.: 384 (1949); K.T.S.: 338 (1961); Vollesen in Opera Bot. 59: 25 (1980); K.T.S.L: 122 (1994). Types: Tanzania, Uzaramo District: near Dar es Salaam, Sachsenwald, *Holtz* 356 (B†, syn., EA!, isosyn.), Pugu Hills, *Holtz* 331 (B†, syn.) & Rufiji District: Mafia I., E coast, *Busse* 415 (B†, syn.)

Much branched evergreen shrub or small tree 0.5–3(–6 fide K.T.S.L.), tall; stems pale, not lenticellate. Leaves thin, narrowly elliptic, oblong-elliptic or oblanceolate, 2.3–7.5(–12) cm long, 0.6–2.2(–3.5) cm wide, aristate at the apex, cuneate at the base; margins entire or slightly serrulate above but with fine elongate gland-tipped setae up to 4.5(–5.5) mm long near base or with spaced setae all round margin; petiole 2–3 mm long; stipules linear-triangular, 4 mm long, 1.5 mm wide at base. Flowers solitary; pedicels 1.3–2 cm long, jointed at base or 1 mm above. Sepals ovate-lanceolate, 10–13 mm long, 3.5–6 mm wide, becoming red in fruit but hardly enlarging. Petals yellow, 10–15 mm long, 7–9 mm wide. Anthers 5–6 mm long, dehiscing by apical pores; filaments 4 mm long. Carpels 5; styles 13–14 mm long, very shortly free at apex. Drupelets black, 6 mm long, 3–4 mm wide.

KENYA. Kilifi District: 13 km from Gotani to Bamba, 21 Nov. 1989, *Luke & Robertson* 2128! & Fumbini, 12 Sept. 1936, *Swynnerton* 16! & Kibarani, 12 Sept. 1936, *Swynnerton* 10!
TANZANIA. Pangani District: Msubugwe Forest, 4 Mar. 1956, *Tanner* 2634!; Uzaramo District: Kisiju, Kure Kese Forest Reserve, Sept. 1953, *Semsei* 1356!; Kilwa District: Selous Game Reserve, Nungu [Nunga] Thicket, 18 Jan. 1977, *Vollesen* in MRC 4334!
DISTR. **K** 7; **T** 3, 6, 8; not known elsewhere
HAB. *Brachystegia* woodland and thicket, *Uapaca*, *Maprounea* etc. bushland, *Pterocarpus*, *Pleurostylia*, *Brachystegia*, *Ostryoderris*, *Sclerocarya* forest and intermediate evergreen rain-forest; 15–700 m

NOTE. *Iversen* et al. (East Usambaras, Marimba Forest Reserve, 1 Nov. 1986) has large leaves up to 12 cm long, 3.5 cm wide, is sterile and probably a young plant. It appears to be this species but further material is needed to confirm that it is not a distinct related species. I have seen no young flowers and better material is needed to fill in some gaps in the description.

4. **Ochna thomasiana** *Eng. & Gilg* in E.J. 33: 245 (1903); T.T.C.L.: 385 (1949); U.O.P.Z.: 382 (1949); K.T.S.: 341 (1961); Vollesen in Opera Bot. 59: 25 (1980); Blundell, Wild Fl. E Afr.: 63, t. 442 (1987); Thulin, Fl. Som. 1: 241 (1993); K.T.S.L: 123 (1994). Type: Kenya, Lamu, *Thomas* 192 (B†, holo., K, photo.!, K!, iso.)

Shrub or small tree 1.5–8.4(–9.6) m tall with grey bark and spreading or drooping branches; young branchlets dark purplish, striate, with dense pale lenticels. Leaves coriaceous or subcoriaceous, narrowly to broadly elliptic, ovate-elliptic or obovate-elliptic, 2–15 cm long, 1–5 cm wide, ± acute at apex, the tip being spinous-apiculate, narrowly rounded to slightly subcordate at base, margin with gland tipped setae 1–5 mm long either restricted to base with rest of margin entire or setose all round; lateral veins ± 20 with very close reticulation of partly subparallel tertiary veins, prominent above; petiole 1–4 mm long; stipules half-navicular, 6 mm long, 1.5 mm wide, very soon deciduous. Flowers solitary or up to 10 in racemiform inflorescences; pedicels 1.3–2.5 cm long, jointed at base or 3–8.5 mm from base. Sepals elliptic-oblong to lanceolate, 10–13 mm long, 2–5 mm wide becoming red in fruit and up to 15 mm long, 7 mm wide. Petals bright yellow, obovate or elliptic, with distinct claw, 12–22 mm long, 8–15 mm wide. Anthers orange, 2.5–5 mm long, dehiscing by apical pores; filaments 2–5 mm long. Carpels 8–11; styles shortly free at apex. Drupelets 7–11 or as few as 1 or 2 by abortion, black, ellipsoid, 7–11 mm long, 5–7 mm wide.

© The Board of Trustees of the Royal Botanic Gardens, Kew, 2005

KENYA. Kwale District: between Samburu and MacKinnon Road, near Taru, 3 Sept. 1953, *Drummond & Hemsley* 4154!; Kilifi District: Arabuko Forest, 1927, *Gardner* in F.D. 1428!; Tana River District: 1 km S of Bfunbe, 4 Aug. 1988, *Robertson & Luke* 5317!
TANZANIA. Lushoto District: Luengera Valley, 26 Aug. 1961, *Semsei* 3291!; Pangani District: Kumbamtoni, 12 Oct. 1956, *Tanner* 3168!; Bagamoyo District: Kikoka Forest Reserve, April 1964; *Semsei* 3831!; Zanzibar: Mazizini [Massazini], 24 Nov. 1959, *Faulkner* 2409!
DISTR. **K** 1, 7; **T** 3, 6; **Z**; **P**; Somalia; also cultivated in Congo (Kinshasa) at Eala and in India
HAB. Evergreen thicket and forest on coral rock and cliffs, coastal bushland on sand down to near high water mark; 0–400 m

SYN. *O. ciliata* Lam. var. *hildebrandtii* Engl., P.O.A. C: 273 (1895). Types: Kenya, Mombasa, *Hildebrandt* 1998 (B†, syn.) & Tanzania, Tanga District: Moa, *Holst* 3071 (B†, syn., K!, isosyn.)
　　*O. kirkii* sensu Gilg in E.J. 33: 245 (1903), *non* Oliv.
　　*Polythecium hildebrandtii* (Engl.) Tiegh. in Ann. Sci. Nat. sér. 8, 16: 372 (1932), *non Ochna hildebrantii* (Baill.) O. Ktze (1891)

5. **Ochna kirkii** *Oliv.* in F.T.A. 1: 317 (1868); Sim, For. Fl. Port. E Afr.: 28 (1909); T.T.C.L.: 384 (1949); Robson in F.Z. 2: 232 (1963); Vollesen in Opera Bot. 59: 25 (1980); K.T.S.L.: 122 (1994). Type: Mozambique/Tanzania, R. Ruvuma, 36 km from mouth, *Kirk* s.n. (K!, holo.)

Shrub or small tree 1–5(–6) m tall with rather rough greyish white bark; branchlets brown or dark purplish, usually prominently lenticellate (save on youngest parts). Leaves thin or ± coriaceous, elliptic or oblong-elliptic to narrowly obovate, 5–21 cm long, 2.5–7 cm wide, obtuse to acute, acuminate or rounded at apex, mucronate and aristate at tip, cuneate to cordate at the base, margins entire or with numerous dark-tipped (at least when dry) ± equal spine-like setae up to 2 mm long; lateral veins ± 20, with closely reticulate tertiary venation of many parallel veinlets all prominent on both surfaces, but less so beneath; petiole (2–)4–5(–6) mm long, rather thick. Flowers 3–± 20 in condensed panicles terminating short lateral shoots; pedicels 1–2.5 cm long, jointed (1–)5–9 mm from the base; bracts 8 mm long, 4 mm wide, striate. Sepals elliptic to oblong-elliptic, 10–16 mm long, (4–)7–8 mm wide, rounded at the apex, becoming red and enlarging to 20 mm long, 10 mm wide. Petals bright yellow, obovate to broadly elliptic, 13–25 mm long, (6–)10–18 mm wide. Anthers orange, 3–5.5 mm long, half as long as or equalling the filaments, dehiscing by apical pores. Carpels 8–10(–12); styles free at apex, spreading; stigmas capitate. Drupelets black, cylindric, compressed, 9–10 mm long, 5.5–8 mm wide, inserted near base.

subsp. **kirkii**

Leaves smaller, 3.3–10 cm long, 1.4–4.5 cm wide, more coriaceous when adult, rounded to subacute at the apex, ± rounded, truncate or slightly subcordate at the base, occasionally cuneate, entire or with setae shorter and much less evident; reticulation of ultimate parallel veinlets fairly prominent.

TANZANIA. Rufiji District: Lake Utunge [Utenge], 15 Dec. 1971, *Ludanga* 1358!; Kilwa District: 6 km NNW of Kingupira, *Vollesen* MRC 2583! & 2.3 km N of Nainokwe village on Kilwa road, 15 Oct. 1978, *Magogo & Rose Innes* 370!
DISTR. **T** 6, 8; Mozambique
HAB. Riverine forest and thicket, groundwater forest, *Pterocarpus*, *Brachystegia*, *Acacia*, *Combretum* woodland; 40–125 m

SYN. *O. carvalhi* Engl., P.O.A. C: 273 (1895); Gilg in E.J. 33: 236 (1903); T.T.C.L.: 383 (1949). Type: Mozambique, Mussoril and Cabecceira, *Carvalho* s.n. (B†, holo., COI, iso.)
　　*O.* sp. nov. aff. *O. kirkii* Oliv.; Vollesen in Opera Bot. 59: 25 (1980)

NOTE. I have been unable to separate the material cited by Vollesen from Mozambique material of typical *O. kirkii* with entire leaves annotated by Robson. Vollesen states some specimens have short anthers 3.5–4 mm long, shorter than the filaments and ± cordate leaf-bases and others have anthers 5.5 mm long, longer than the filaments and rounded leaf-bases but I have found filaments 4–7 mm long in one inflorescence.

© The Board of Trustees of the Royal Botanic Gardens, Kew, 2005

subsp. **multisetosa** Verdc. **subsp. nov.** a subsp. *kirkii* foliis majoribus, 5–21 cm longis, 2.5–7 cm latis, tenuioribus, plerumque apice acutis vel acuminatis, basi cuneatis, margine valde setosis, venulis ultimis parallelis valde prominentibus differt. Typus: Tanzania, Lushoto District: East Usambaras, 2.4 km N of Mpandeni [Pandeni] on track to Longusa, *Drummond & Hemsley* 3481 (K!, holo. & iso.)

Leaves larger, 5–21 cm long, 2.5–7 cm wide, thinner when adult, mostly acute or acuminate at the apex, cuneate at the base, with very evident close setae at margins; reticulation of ultimate parallel veinlets very evident.

Kenya. Kwale District: 24 km SW of Kwale, Mwasangombe Forest, 27 Aug. 1953, *Drummond & Hemsley* 4013!; Lamu District: Mambosasa, Utwani Forest Reserve, 18 Oct. 1957, *Greenway & Rawlins* 9367! & Dec. 1956, *Rawlins* 222!

Tanzania. Lushoto District: Sigi Gorge, 25 Aug. 1942, *Greenway* 6623!; Uzaramo District: Pugu Forest Reserve, June 1954, *Semsei* 1728! & Pugu Hills, above St. Andrews, Thinaki, 5 Aug. 1969, *Batty* 584!; Pemba, Vitongoje, 4 Aug. 1929, *Vaughan* 432 (fide EA)

Distr. **K** 7; **T** 3, 6; **P**; not known elsewhere

Hab. Semi-deciduous forest of *Gyrocarpus*, *Antiaris*, *Milicia*, *Afzelia*, *Parkia* etc., groundwater forest, rain-forest of *Pterocarpus*, *Combretum schumannii*, *Cynometra*, *Antiaris* etc., also in *Hyphaene* grassland and on exposed rock-faces; 100–600 m

Syn. *O. thomasiana* sensu Engl. & Gilg in E.J. 33: 245 (1903) pro parte, *non* Engl. & Gilg sensu stricto

    *O.* sp.; T.T.C.L.: 284 (1949) adnot. quoad *Greenway* 4639 & 4688

    *O.* sp.; K.T.S.: 341 adnot. (1961)

Note. *Semsei* 3348 (Tanzania, Kilosa, 25 Oct. 1961) said to be a small tree 4.5 m tall has the setiform marginal teeth much less pronounced and is doubtfully included here. It could be a form of *O. macrocalyx* or possibly a hybrid but no hybrids have been reported for the genus. *Kirk* 132, Dar es Salaam, 1 Nov. 1869, seems to lack the marginal teeth and has been determined as *O. kirkii* by Robson.

6. **Ochna mossambicensis** *Klotzsch* in Peters, Reise Mossamb. Bot. 1: 88, t. 16 (1861); Oliv. in F.T.A. 1: 317 (1868); Gilg in E.J. 33: 234, 244 (1903); Sim, For. Fl. Port. E Afr.: 28 (1909); T.T.C.L.: 384 (1949); K.T.S.: 340 (1961); Robson in F.Z. 2: 233 (1963); Vollesen in Opera Bot. 59: 25 (1980); Blundell, Wild Fl. E Afr.: 63, t. 277 (1987); Thulin, Fl. Somalia 1: 241, fig. 133 (1993); K.T.S.L.: 122 (1994). Type: Mozambique, Sena, *Peters* s.n. (B†, holo., EA!, K!, iso.)

Shrub, small tree or rhizomatous subshrub, 0.05–5(–9) m tall; bark pale grey or brown, smooth or rather rough and fissured; branches rather thick. Leaves coriaceous, obovate to oblanceolate or oblong, (3.5–)5.5–22.5 cm long, (1.5–)2–8.4 cm wide, obtuse to broadly rounded at the apex, less often subacute with a short mucro, cuneate to subtruncate at the base, margin densely serrulate; lateral veins 20–25, prominent on both surfaces; tertiary venation finely reticulate, more prominent above than beneath; petiole ± stout, 1.5–8 mm long. Flowers numerous in branched panicles terminating lateral shoots; pedicels 1.5–3 cm long, jointed 4–8 mm from base, the joints forming characteristic tufts when flowers have fallen. Sepals elliptic to oblong-elliptic, 9–11(–12.5) mm long, 5–8 mm wide, rounded at apex, becoming scarlet-red and slightly enlarging to 12–14 mm long in fruit. Petals bright yellow, obovate or almost round, 10–22 mm long, 7–19 mm wide. Anthers orange-yellow, 4–7 mm long, 3–5 times as long as the filaments, dehiscing by apical pores. Carpels (6–)8–10; styles united almost to apex, the free ends spreading; stigmas slightly enlarged. Drupelets black, subglobose or ovoid-cylindric, 8–10 mm long, 6–8 mm wide, inserted near the base.

Kenya. Kwale District: 56 km from Mombasa on main Nairobi road, 12 Aug. 1959, *Verdcourt* 2347!; Mombasa District: Nyali Bridge, mainland, 26 Jan. 1953, *Drummond & Hemsley* 1013!; Tana River District: 48 km S of Garsen, 5 Oct. 1961, *Polhill & Paulo* 592!

Tanzania. Tanga District: Sawa, 20 June 1956, *Faulkner* 1889!; Usaramo District: Kiserawe Forest Reserve, Aug. 1953, *Semsei* 1333!; Lindi District: Rondo Forest Reserve, 21 Aug. 1967, *Shabani* 34!; Zanzibar: Marahubi, Feb. 1930, *Vaughan* 1277!

© The Board of Trustees of the Royal Botanic Gardens, Kew, 2005

DISTR. **K** 1, 7; **T** 3, 6, 8, **Z**; Somalia, N Mozambique

HAB. Various mixed thicket and bushland e.g. *Commiphora – Cassia, Acacia – Commiphora – Dichrostachys – Hibiscus*; evergreen forest of *Combretum schumannii, Trachylobium, Afzelia, Dichrostachys, Brachystegia, Cynometra* etc., sometimes on cliffs within spray zone of high tide and on shore just above high water mark; 0–450 m

SYN. *O. fisheri* Engl. in E.J. 17: 78 (1893). Type: Tanzania, Handeni District: Wadiboma, *Fischer* 92 (B†, holo.)

    *O. purpureo-costata* Engl., P.O.A. C: 273 (1895). Type: Tanzania, Uzaramo District: Useguha, *Stuhlmann* 7082 (B†, holo.)*

NOTE. The only record from **K** 1 is *Joy Adamson* 138 in *Bally* 6038 from Galma Galla, 3 Sept. 1945 having much smaller leaves. Robson has annotated it as a forma. More material is needed to determine its status. A specimen at the BM, *Salt* s.n., 'Abyssinia' has been determined by Robson as *O. mossambicensis* but is not mentioned by Vollesen.

7. **Ochna macrocalyx** *Oliv.* in F.T.A. 1: 319 (1868) & in Trans. Linn. Soc. 29: 43, t. 19 (1873); Gilg in E.J. 33: 236 (1903); Sim, For. Fl. Port. E Afr.: 28 (1909); T.T.C.L.: 384 (1949); Robson in F.Z. 2: 234, t. 44, fig. A (1963); Haerdi in Acta Trop. Suppl. 8: 100 (1964); Vollesen in Opera Bot. 59: 25 (1980). Types: Tanzania, District unclear, Mbumi, 6°56' S, *Grant* s.n.; Malawi, Soche Mts [Sotschi], *Kirk* s.n. & ?Mozambique, Manganja Mts, *Meller* s.n. (all K!, syn.)

Suffrutex or small shrub 0.1–3 m tall, sometimes with only leaves and flowers showing above ground; bark greybrown, rough; branchlets purplish, usually lenticellate, later greyish with epidermis sometimes peeling. Leaves coriaceous, oblong- to oblanceolate-elliptic, 6–19(–25) cm long, 1.6–4.5(–7) cm wide, acute to obtuse at the apex, cuneate to rounded at the base, margin densely, evenly and acutely serrulate; lateral veins and tertiary reticulate venation prominent above, less so beneath; petiole 1–3 mm long. Flowers precocious or occurring with leaves, (2–)3–9(–14) in axillary receme-like inflorescences with short rhachis; pedicels 1.3–4 cm long with joint at 2–5 mm from base. Sepals elliptic to narrowly ovate, 1.2–2.2(–2.5) cm long, 0.6–1 cm wide, rounded at the apex, becoming red and attaining 2–3 cm long, 0.8–1.5 cm wide in fruit. Petals bright yellow or orange-yellow, elliptic, obovate or almost round, 1.6–3.2 cm long, 0.9–2.7 cm wide. Anthers (4.5–)7–8 mm long, as long as or twice as long as the filaments, dehiscing by apical pores. Carpels 5(–10); styles free at apex and recurved with capitate stigmas. Drupelets black, oblong-ellipsoid or subcylindrical, 9–13 mm long, 5–9 mm wide, inserted near base. Fig. 1 (p. 13).

TANZANIA. Lushoto District: Kijango, 27 Oct. 1935, *Greenway* 4136!; Mpanda District: plateau above Ngungusi escarpment, 24 Oct. 1959, *Richards* 11536!; Songea District: Gumbiro, 24 Jan. 1956, *Milne-Redhead & Taylor* 8515!

DISTR. **K** 7 (see note); **T** 1, 3–8; Zambia, Malawi, Mozambique, Zimbabwe

HAB. Open *Combretum* and mixed *Brachystegia, Combretum, Pterocarpus, Julbernardia, Isoberlinia* woodland, sometimes on rocky slopes, grassland, cracks in rock-faces, also in cultivations derived from woodland; 350–1350 m

SYN. *O. macrocarpa* Engl. in E.J. 17: 77 (1893); P.O.A. C: 273 (1895); Gilg in E.J. 33: 236 (1903). Type: Tanzania, Shinyanga District: Usukuma, between Usulu and Usiha, *Fischer* 90 (B†, holo.)

    *O. splendida* Engl. in E.J. 28: 434 (1900) & in E.J. 30: 355, fig. A–F & 356; Gilg in E.J. 33: 236 (1903). Type: Tanzania, Morogoro District: Uluguru, between Mgeta and Mbakana, *Goetze* 335 (B†, holo., BR!, K!, iso., EA, photo.)

NOTE. Robson (F.Z. 2: 236 (1963)) states "the number of carpels in our area [F.Z.] is apparently always 5. In a closely related species of fringing forest and rain-forest margins in Morogoro District, Tanzania, however, it varies from 6–10 and specimens of *O. macrocalyx* from that area

* Engler cites two distinct localities but only one specimen number.

© The Board of Trustees of the Royal Botanic Gardens, Kew, 2005

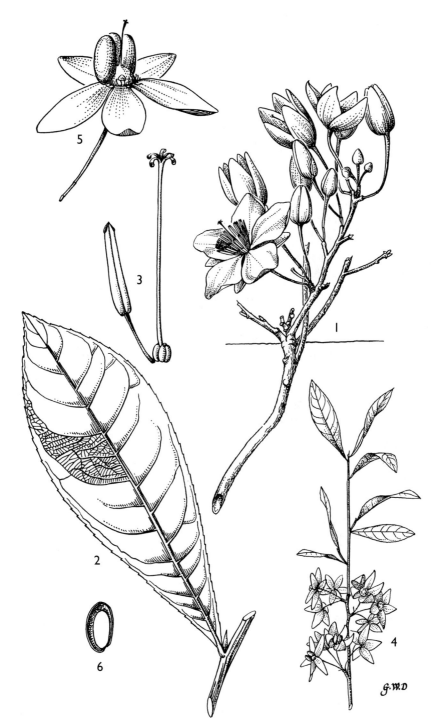

FIG. 1.  *OCHNA MACROCALYX* — **1**, flowering shoot of dwarf form, × 1; **2**, leaf, × 1; **3**, stamen and gynoecium, × 4; **4**, fruiting branch of tall form, × ¹/₄; **5**, fruit, × 1; **6**, section of drupelet, × 1. 1, 3 from *Robson* 659; 2 from *Faulkner* P21; 4–6 from from *Carson* s.n. Drawn by G.W. Dalby, and reproduced from Flora Zambesiaca.

© The Board of Trustees of the Royal Botanic Gardens, Kew, 2005

sometimes have 6–8". Vollesen in Opera Bot. 59: 25 (1980) mentions *O.* sp. nov. aff. *O. macrocalyx* which differs in habit and in the larger number of carpels but is otherwise very close. He cites *Carmichael* 147, *Burtt* 5405, *MRC* 537 and *Vollesen MRC* 4062. I have been unable to sort out this variable assemblage and am treating all as *O. macrocalyx*. Dr. Wadhwa during preliminary studies had reached the same conclusion. Doubt remains and field studies on correlation of habit and carpel number with habitat are needed.

The **K** 7 indication above is based on *Luke & Mbinda* 5847 from Kwale District: Shimoni collected in coral rag forest at 10 m on 11 May 1999. It is a tree to 5 m with rather thinner leaves with rather more rounded base and more slender pedicels; the bright red fruiting sepals are up to 3 × 1.5 cm. Further flowering material is needed to confirm its identity (see also note after *O. insculpta*).

8. **Ochna insculpta** *Sleumer* in N.B.G.B. 12: 68 (1934); K.T.S.: 338, fig. 66 (1961); Blundell, Wild Fl. E Afr.: 63 (1987); K.T.S.L: 122 (1994); Vollesen in Fl. Ethiopia & Eritrea 2 (2): 67, fig. 69. 2. 1 & 2 (1995). Type: Kenya, Embu District: E Mt Kenya, Kirimiri, *R.E. & T.C.E. Fries* 2028 (UPS!, holo., B†, iso., UPS!, iso.)

Shrub 1–2.5 m tall or tree 7–9 m tall with smooth brown bark or sometimes rough and finely fissured; young branchlets angular or ridged, densely lenticellate; flowering branchlets often leafless. Leaves very glossy and bronze-flushed when young, thin, oblong-elliptic, 2.5–12(–14) cm long, 1.2–4.3(–5) cm wide, acute, slightly acuminate or narrowed to a ± rounded apex, cuneate to ± rounded at the base, spinulose-serrate or ± setose at the margin, the spinules subulate, facing apex; lateral veins 13–15, prominent, but the actual area of lamina around the veins very often conspicuously impressed; tertiary venation reticulate, very prominent above, less so beneath; petiole 1–3 mm long; stipules narrowly triangular, 6 mm long, 3 mm wide, serrate near tip. Flowers 1–6 in short raceme-like or subumbellate inflorescences, the rhachis 0–7 mm long or flowers solitary; pedicels 2–3.3 mm long, jointed (1–)5–7 mm from the base. Sepals ovate-oblong, 1.2–2.2 cm long, 6–8 mm wide, becoming red but scarcely enlarging in fruit up to 2.4 cm long, 1 cm wide. Petals yellow, round or elliptic, (8–)18–25 mm long, (9–)13–18 mm wide, narrowed at the base. Anthers 4–6.5 mm long, slightly longer than the filaments, opening by apical pores. Carpels 5–7; styles very shortly branched at apex. Drupelets black, ellipsoid or subcylindrical, 10–12 mm long, 7–8 mm wide (rarely 8 × 4.5 mm). Fig. 2 (p. 15).

UGANDA. Mengo District: Buruma Is., 20 Mar. 1904, *Bagshawe* 636! & km 21 on Entebbe road [from Kampala], Nov. 1937, *Chandler* 2031!
KENYA. Northern Frontier District: Marsabit, 14 Feb. 1953, *Gillett* 15113!; Nairobi District: Karura Forest, 30 Mar. 1947, *Bogdan* 468!; Kericho District: Sotik, Nandaret Estate, Oct. 1957, *Dale* 1024!
TANZANIA. Moshi District: Old Moshi, *Semkiwa* 78! & Arusha District: Ngurdoto National Park, Longil, 28 Feb. 1966, *Greenway & Kanuri* 12400 ! & same area, Lochu, 1 Mar. 1966, *Greenway & Kanuri* 12407!
DISTR. **U** 4; **K** 1, 3–6; **T** 2; Ethiopia
HAB. Evergreen forest e.g. *Olea – Turraea – Croton* and *Juniperus – Podocarpus – Calodendrum*, riverine forest and particularly in thicket and woodland at forest edges; 1050–2100 m

SYN. *O. nandiensis* Dale, Trees & Shrubs Kenya Colony: 27 (1936) *nomen anglice,* Jex-Blake, Gard. E Afr. ed. 4: 267 (1957) (name very widely used but never validated)
    *O. boranensis* Cufod. in Miss. Biol. Borana Racc.-Bot.: 138 (1939). Type: Ethiopia, Sidamo, *Cufodontis* 291 & 353 (FT, syn.)
    *O.* sp.; Jex-Blake, Gard. E Afr. ed. 3, 208, t. 1 (1949)
    *O.* sp. near *macrocalyx,* I.T.U. ed. 2: 281 (1952)

NOTE. Gillett sent a note to the botanist 'in charge of Ochnaceae' to be put in the covers (dated 17 May 1979) discussing the relation between *O. insculpta* and *O. macrocalyx*. He thought these could just be distinguished using size of plant, length of fruiting calyx and ecological preferences. Until the taxonomy of the variable *O. macrocalyx* is better understood

© The Board of Trustees of the Royal Botanic Gardens, Kew, 2005

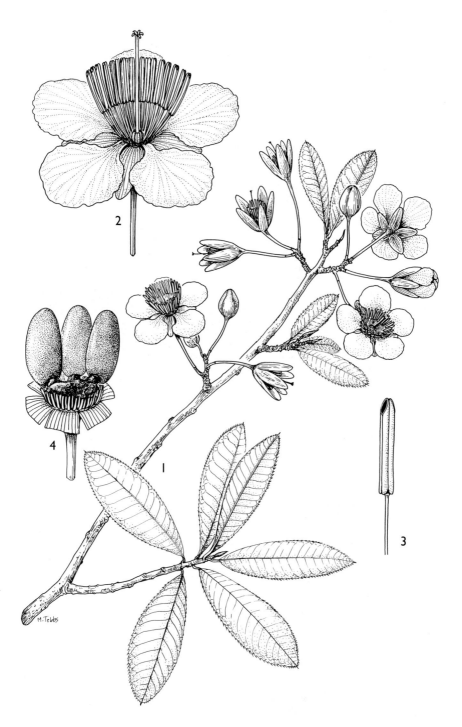

FIG. 2 *OCHNA INSCULPTA* — **1**, habit, × ²/₃; **2**, flower, × 2; **3**, stamen, × 4; **4**, fruits, several drupelets removed, × 2. 1 from *Williams* 327; 2–3 from *Bally & Smith* 14809; 4 from *Hepper & Jaeger* 6969b. Drawn by Margaret Tebbs.

© The Board of Trustees of the Royal Botanic Gardens, Kew, 2005

*O. insculpta* is best kept distinct. It has been cultivated at various places in Nairobi but material in the Arboretum is probably part of the original cover or derived from it. *Gardner* in F.D. 1402 states timber tree but his specimen was collected from coppice shoots in the Nature Reserve outside the forest. I think this and other reports of *O. inculpta* becoming a large tree are due to confusion with *O. holstii.*

9. **Ochna polyarthra** *Verdc.* **sp. nov.** in Sect. *Ochna* ponenda affinis *O. massambicensi* Klotzsch confinibusque combinatione inflorescentiarum condensarum 30–40-florarum cum petalis parris 9–10 mm longis 4–5.5 latis carpellis undecim distinguitur. Typus: Tanzania, Uzaramo District: Kisarawe Forest Reserve, *Mgaza* 707 (K!, holo., EA, iso.)

Shrub about 3.6 m tall with grey thick branches with bark ± ridged and peeling; youngest parts without obvious lenticels. Leaves not adult at flowering stage, restricted to youngest part of stems, narrowly oblong-oblanceolate, about 9 cm long, 2.3 cm wide, probably ± acute at apex, cuneate at the base, margins with rather sparse aculeate teeth, venation not yet developed and prominent save at extreme base; petiole ± 8 mm long with leaf-base narrowly decurrent; stipules triangular, up to 6 mm long, striate. Flowers on leafless branches on spur shoots so closely placed that they form a tight 30–40-flowered 3.5 × 4.5 cm cluster at apices of branches; pedicels slender, 17–25 mm long, jointed ± 5 mm from the base. Sepals elliptic-oblong, 8–9 mm long, 2.5–4 mm wide. Petals yellow, elliptic, 9–10 mm long, 4–5.5 mm wide, unguiculate. Anthers slender, 6 mm long, opening by small apical pores, filaments 2.5 mm long. Carpels 11; stigma capitate, ± lobed. Fruits not known.

TANZANIA. Uzaramo District: Kisarawe Forest Reserve, 16 Oct. 1965, *Mgaza* 707!
DISTR. **T** 6; known only from the type
HAB. ?Dry evergreen forest; ? under 200 m

NOTE. The specific epithet is a plural noun in apposition.

10. **Ochna citrina** *Gilg* in E.J. 33: 246 (1903); T.T.C.L.: 383 (1949). Type: Tanzania, Lindi District, near Mtama, *Busse* 1114 (B†, holo.)

Subshrub 30–35 cm tall with smooth brown branches. Leaves coriaceous, obovate, 13–15 cm long, 6–8 cm wide, ± rounded at the apex, narrowly cuneate at the base, regularly spinulose-serrate, shiny on both sides; lateral veins 15–20, ± at right angles to midrib together with very numerous transverse tertiary veinlets, very prominent above, less so beneath; petiole 2–3 mm long and thick. Flowers few at apices of branches, often forming paniculiform clusters; pedicels ± 1 cm long. Sepals oblong, 11–12 mm long, 4–5 mm wide, rounded at apex. Petals lemon yellow, larger than the sepals, broadly obovate, 14–15 mm long, 8–9 mm wide, rounded at the apex, narrowed at base. Anthers linear, 5 mm long, opening by apical pores; filaments shorter, 3–4 mm long. Style 11–12 mm long, divided at apex into numerous very short stigma-bearing branches. Fruits not seen.

TANZANIA. Lindi District: near Mtama, ± 5 Mar. 1901, *Busse* 1114
DISTR. **T** 8; known only from the type (destroyed)
HAB. Sandy places in open *Brachystegia* woodland; ± 200 m

11. **Ochna braunii** *Sleumer* in F.R. 39: 276 (1938); T.T.C.L.: 383 (1949). Type: Tanzania, Lindi District: Rondo–Lutamba, *Braun* 1209 (B†, holo., EA!, iso.)

Shrub 1–2 m tall; branches with rugose grey-brown bark; young shoots purplish and lenticellate. Leaves lanceolate to elliptic-lanceolate, rigidly chartaceous, 5.5–8(–10.5) cm long, 1.2–2.5 cm wide, acute at apex, widest at the middle, attenuate

© The Board of Trustees of the Royal Botanic Gardens, Kew, 2005

at base, margin densely regularly serrate-dentate, the teeth 0.5 mm long, 2–3 mm apart; midrib prominent on both surfaces; other venation obscure above but minutely prominent beneath; petiole 4 mm long. Flowers 2–9 in ± abbreviated axillary cymes towards the end of the lateral branches; pedicels slender, 1–1.5 cm long in flowering state but eventually elongating to 2.3 cm, jointed at the base. Sepals red, elongate-ovate, 1 cm long, 4 mm wide. Sepals oblong-elliptic, up to 1.5 cm long, 5 mm wide. Petals oblong-spathulate, 13–20 mm long, 7–13 mm wide. Anthers 6–7 mm long, opening by apical pores; filaments 1.5–4 mm long. Styles 8–13 mm long, free at the apex. Drupelets 1–2, black, ellipsoid, 8 mm long, 5–5.5 mm wide.

TANZANIA. Lindi District: Rondo-Lutamba, 13 June 1906, *Braun* 1209! & Rondo Plateau [Mwera Plateau], ± 50 km W of Lindi, *Schlieben* 6530 & Rondo Plateau, St. Ciprian's College, 2 June 2001, *Anthony* 26!
DISTR. **T** 8; not known elsewhere
HAB. Semi-decidous thicket with *Brachylaena* on plateau edge, rarely in woodland on plateau; 500 m

NOTE. Sleumer states this species resembles *O. kirkii* Oliv. but can be distinguished from it by the absence of a long drawn-out leaf apex. At first I was not satisfied with the taxonomy of this species. Study of the EA isotype, which Sleumer would not have seen, reveals a number of discrepancies when compared with Sleumer's description. The EA isotype has no young flowers and it is almost certain the Berlin holotype would have had none either in which case anther, petal and young sepal information would have been derived from *Schlieben* 6530 of which I have been unable to find a duplicate. The persistent filaments in *Braun* 1209 are 4 mm long not 1.5 mm and the persistent styles 12 mm not 8. Brother J. Anthony's flowering specimen shows this is within the variation possible in this species. *Braun* 1209 had been first identified as *Ochna holstii* by some German worker at Amani against which Greenway has written a firm 'not'. He was of course familiar with the tall timber tree in the East Usambaras but Robson has accepted shrubby forms.

12. **Ochna schliebenii** *Sleumer* in F.R. 39: 278 (1936); T.T.C.L.: 385 (1949). Type: Tanzania, Lindi District: 20 km S of Lindi, *Schlieben* 5777 (B†, holo., BM!, BR!, iso.)

Small shrub with red-brown branches. Leaves drying dark olive-brown above, oblanceolate, 9–16 cm long, 3–4(–5) cm wide, acuminate to a subobtuse apex, attenuate to long-cuneate at the base, margin irregularly serrate-dentate, the teeth very acute, hardly 1 mm long, (2–)3–4 mm apart; midrib very prominent beneath; lateral veins rather irregular, the main ones 1–1.5 cm apart, arcuate-ascending with numerous intermediaries, together with tertiary venation very prominent above but much less so beneath; petiole thick, 2–3 mm long. Flowers in few-flowered lax very shortly pedunculate cyme-like inflorescences; pedicels ± 1.5 cm long, jointed 3–4 mm from base. Sepals reddish, oblong, 8 mm long, 3 mm wide in flower but ovate-triangular in fruit and attaining 1.7 cm long, 6 mm wide. Petals yellow, obovate-oblong, 13 mm long, 7 mm wide, distinctly unguiculate. Anthers 5.5–7 mm long, dehiscing by apical pores, subsagittate at base; filaments 1–2 mm long. Drupelets in fruits seen single, black, subglobose, 8 mm diameter.

TANZANIA. Lindi District: 20 km S of Lindi, Mlinguru, 22 Dec. 1934, *Schlieben* 5777!
DISTR. **T** 8; known only from the type
HAB. Bushland; ± 270 m

13. **Ochna bracteosa** *Robyns & Lawalrée* in B.J.B.B. 18: 278 (1947); Robyns, F.P.N.A. 1: 614 (1948); I.T.U. ed. 2: 279 (1952); Bamps, F.C.B. Ochnaceae: 20 (1967); Vollesen in Fl. Ethiopia & Eritrea 2 (2): 67, fig. 69.1.7 (1995). Type: Congo (Kinshasa), Parc National Albert, Lesse, *Bequaert* 3198 (BR, holo.)

© The Board of Trustees of the Royal Botanic Gardens, Kew, 2005

Small forest undershrub 0.3–2.4 m tall; branchlets grey or brownish to purplish, lenticellate; buds fusiform, 4–5 mm long, 1.5 mm diameter, the scales entire and persistent on the young stems, ± imbricate, lanceolate, 2–5 mm long, 1–1.5 mm wide. Leaves thin, oblanceolate to elliptic, (3.5–)7–11 cm long, (1–)2–3.5 cm wide, acuminate to an acute apex, cuneate at the base, margin serrulate; lateral veins and somewhat sparse tertiary venation finely raised on both surfaces; petiole 1–4 mm long; stipules linear-lanceolate, 4–5 mm long, 1 mm wide, entire, soon deciduous. Flowers solitary; pedicel 0.5–1(–1.5) cm long, jointed 2.5–4 mm above base. Sepals oblong-elliptic, (6–)8–10 mm long, 3–5 mm wide, becoming red in fruit, 10–14 mm long, 5–6 mm wide. Petals yellow, elliptic-oblanceolate, 9–13 mm long, 3–4 mm wide. Anthers 12–16, 4 mm long, longer than the filaments, opening by apical pores. Carpels 3–4; style shortly 3–4-fid at apex. Drupelets black, ellipsoid, 7–8 mm long, 5–6 mm wide.

UGANDA. Bunyoro District: Budongo Forest, Nov. 1932, *Harris* 152 in F.D. 1116!; Toro District: Bwamba, Nabulongwe Forest, 19 Dec. 1949, *Dawkins* 480!; Mengo District: Mabira Forest, Oct. 1922, *Dummer* 5565!
DISTR. **U** 2–4; Cameroon, E Congo (Kinshasa), Sudan, Ethiopia
HAB. Understorey of evergreen forest; 800–1200 m

14. **Ochna inermis** (*Forssk.*) *Schweinf.* in Penzig in Atti Congr. Bot. Intern. Genova 1892: 335 (1893) & Arab. Pfl. Aegypt. Alger. & Jemen: 148 (1912); I.T.U. ed. 2: 279 (1952); K.T.S.: 338 (1961); Robson in F.Z. 2: 237 (1963); du Toit & Obermayer, F.S.A. 22: 10 (1976); Blundell, Wild Fl. E Afr.: 63, t. 498 (1987); Thulin, Fl. Somalia 1: 243 (1993); Hepper & Friis, Pl. Forssk. Fl. Aegypt-Arab: 207 (1994); K.T.S.L.: 122 (1994); Vollesen, Fl. Ethiopia & Eritrea 2 (2): 67, fig. 69.1, 5 & 6 (1995); Wood, Handb. Yemen Fl.: 97, t. 3 (1997). Type: Yemen, Al Hadiyah [Hadîe], between Ersch and Alûdje, *Forsskål* 760 (C, holo.)

Deciduous shrub or small tree 1–4.5(–6) m tall with smooth whitish or dark grey bark, extensively branched, the young branches brown or blackish purple at first, later becoming whitish, lenticellate. Leaves reddish or bronze when young, fairly thin to subcoriaceous, elliptic, obovate, oblong or ± round, (1.4–)1.7–4.8(–7.5)(–15 fide Vollesen) cm long, 0.9–2.8(–3)(–7.5 fide Vollesen) cm wide, rounded or obtuse at the apex (rarely ± acute), cuneate to truncate or subcordate at the base, margins sharply curved-serrate; lateral veins together with dense reticulate tertiary venation prominent on both surfaces; petiole 1–2.5(–4) mm long; stipules linear-lanceolate, ± 5 mm long, fimbriate at apex. Flowers precocious or with very young leaves, 1(–2) (rarely 4), on short node-like shoots often closely placed on branchlets; pedicels (0.5–)1.3–3 cm long, jointed (0.5–)1–3(–4) mm from the base; young buds depressed globose, 2–3 mm diameter. Sepals green or orange, elliptic, 5–6 mm long, 4 mm wide, rounded, becoming crimson-red in fruit and enlarging to 8–16 mm long, 7–8 mm wide, convex, usually remaining imbricate around the fruit. Petals bright yellow, obovate or round, 7–11 mm long, 4.5–6 mm wide. Anthers orange-yellow, 1–1.5 mm long, $\frac{1}{2}$–$\frac{1}{3}$ as long as the filaments, dehiscing by apical pores. Carpels 5; styles united almost to apex, recurved at tips. Drupelets black, ellipsoid, 7–10 mm long, 5–7 mm diameter.

UGANDA. Karamoja District: Moroto, Feb. 1936, *Eggeling* 2965! & Matheniko, Lokitanyala [Lokitaungyala], May 1954, *Philip* 588! & near Koputh, *Brasnett* 173!
KENYA. Northern Frontier District: Dandu, 3 Apr. 1952 (fl.), 9 May 1952 (fr.), *Gillett* 12675!; West Suk District: N of Marich Pass, at foot of Kaimat escarpment, 27 Oct. 1977, *Carter & Stannard* 70!; Machakos District: 217 km from Mombasa on Nairobi road, near Kenani, 30 Aug. 1959, *Verdcourt* 2392!
TANZANIA. Kilosa District: Ruaha R. Gorge, 8 Jan 1975, *Brummitt & Polhill* 13613! & Elphon's Pass, 24 Jan. 1988, *Lovett & Congdon* 2956!; Iringa District: Ruaha National Park road turn-off, km 69 on the Iringa–Idodi road, 27 Nov. 1970, *Greenway & Kanuri* 14684!

© The Board of Trustees of the Royal Botanic Gardens, Kew, 2005

DISTR. **U** 1; **K** 1, 2, 3 (fide EA), 4, 6 (fide EA), 7; **T** 6, 7; Eritrea, Ethiopia, Somalia, Mozambique, Zimbabwe, Botswana, N South Africa; Yemen
HAB. Mixed bushland and scrub of *Acacia, Commiphora, Sterculia, Ficus, Cordia, Delonix* etc., often in rocky places, woodland; (100–)350–1500 m

SYN. *Euonymus inermis* Forssk., Fl. Aegypt-Arab. 204 (1775)
   *O. monantha* sensu I.T.U. ed. 2: 280 (1952) quoad *Eggeling* 2965, *non* Gilg?
   *O. ovata* sensu I.T.U. ed. 2: 280 (1952) quoad *Brasnett* 173, *non* F. Hoffm.

NOTE. *Bidgood et al.* 898 from Tanzania, Kilosa District: Malolo–Kisanga track, 3 Apr. 1988 has been annotated as a sp. nov. aff. *O. inermis* but I see no reason for this and agree with Wadhwa's determination as *O. inermis.*

15. **Ochna monantha** *Gilg* in E.J. 33: 247 (1903); T.T.C.L.: 384 (1949). Types: Tanzania, ?Moshi District: 'Massaisteppe' between Kilimanjaro and Meru, *Merker* s.n. & Iringa District: Uhehe, Lukosse R., *Goetze* 476 (both B†, syn.)

Divaricately-branched shrub up to 1 m tall; stems grey, very densely lenticellate. Leaves immature at flowering time, elliptic, 1.8–2.3 cm long, 1.1–1.4 cm wide, ± rounded at the apex and base, obsoletely regular-serrulate at the margin; petiole ± 2 mm long. Flowers solitary at ends of short twigs; pedicels 1.3–1.4 cm long. Sepals ovate, 5.5 mm long, 3–4 mm wide, rounded. Petals yellow, broadly obovate, 11–12 mm long, 8–9 mm wide, rounded or retuse at the apex, narrowly long-unguiculate at the base. Anthers oblong, 1–1.2 mm long, dehiscing by apical pores; filaments 3 mm long. Styles joined, 4 mm long, free for a short distance at apex. Ovary 5-lobed; fruits not seen.

TANZANIA. Moshi District: between Kilimanjaro and Meru, *Merker* s.n.; Iringa District: Uhehe, Lukosse R., 13 Jan. 1899, *Goetze* 476
DISTR. **T** 2, 7; not known elsewhere (but see Note)
HAB. Dry bushland ('Steppe'); ± 800 m

SYN. *O. atropurpurea* sensu Engl. quoad *Goetze* 476 in E.J. 30: 356 (1901), *non* DC.

NOTE. I think this is probably *O. inermis* (Forssk.) Schweinf. but it could be a form of *O. ovata* F. Hoffm.; material cited as *O. monantha* in I.T.U. ed. 2: 280 (1952) is *O. hackarsii* Robyns & Lawalrée.

16. **Ochna ovata** *F. Hoffm.*, Beitr. Kenntn. Fl. Centr. Ost-Afr.: 19 (1889); P.O.A. C: 272 (1895); Gilg in E.J. 33: 234 (1903); T.T.C.L.: 384 (1949); Wimbush, Cat. Kenya Timbers: 57 (1957); K.T.S.: 340 (1961); Vollesen in Opera Bot. 59: 25 (1980); Blundell, Wild Fl. E Afr.: 83, t. 278 (1987); K.T.S.L.: 122, fig. (1994). Type: Tanzania, District unclear, Ugalla R., *Boehm* 342 (B†, holo.)

Deciduous shrub or small tree 1.8–9 m tall with dark brown, grey or ± black rough bark; branchlets purplish grey, often ± densely lenticellate. Leaves coppery when young, thin, narrowly to broadly elliptic, elliptic-oblong or lanceolate-elliptic, 1.3–6(–8) cm long, 0.8–2.5(–4) cm wide, acute to obtuse at the apex, rounded to subcordate at the base, margins closely evenly serrate; venation rather pale, prominently reticulate on both surfaces; stipules linear-lanceolate, 4 mm long, usually fimbriate or undivided save at extreme apex, ± persistent until the leaves thicken; petiole 1–3 mm long. Buds fusiform, 7 mm long, 2.5 mm wide, bracts often adhering to pedicel bases. Inflorescences forming paniculate clusters up to 5 cm diameter, the side branches reduced to short 1–3(–4)-flowered nodules, usually precocious, often flowering when quite leafless; pedicels 1.6–2.5 cm long, jointed 3–6 mm from base. Sepals oblong, ± unequal, 6–8 mm long, 2.5 mm wide, obtuse, becoming scarlet and enlarging to 9–10 × 3.5–5 mm. Petals yellow, obovate, 12–14 mm long, 6–7 mm wide, veined. Anthers 1.5–2 mm long, dehiscing by apical pores; filaments 2–5 mm long. Carpels 5; styles joined save at apex, the branches short and slightly spreading. Drupelets black, ellipsoid 7–9 mm long, 5–6 mm wide, attached near base.

© The Board of Trustees of the Royal Botanic Gardens, Kew, 2005

UGANDA. Karamoja District: Pian County, Lodoketeminit, 6 Apr. 1962, *Kerfoot* 3695! & 20 Feb. 1963, *Kerfoot* 4762!

KENYA. Nairobi District: Kileleshwa road, by Andueni R., 4 Feb. 1952, *G. Williams Sangai* 324!; Masai District: Ngong Escarpment, 21 Dec. 1947, *Bally* 5737!, 5738!; Teita District: Wusi to Mwatate road, 18 Sept. 1953, *Drummond & Hemsley* 4411!

TANZANIA. Tabora, *C.H.N. Jackson* 15!; Dodoma District: Manyoni, Hika village, 10 Dec. 1931, *B.D. Burtt* 3502!; track to Trekamboga Rapids, 12 Dec. 1970, *Greenway & Kanuri* 14789! & 14785!

DISTR. U 1; K 1–7 (2, 3, 5 fide EA); T 1–8; apparently not recorded elsewhere

HAB. Woodland of various types: *Calodendrum – Croton – Albizia, Commiphora – Turraea, Albizia, Gyrocarpus, Cordyla, Isoberlinia – Brachystegia – Burttia*, particularly in Itigi type thicket on rocky hills, dry evergreen forest; sometimes riverine; (500–)800–1800(–2100) m

SYN. *O. stuhlmannii* Engl. in E.J. 17: 77 (1893); P.O.A. C: 273 (1895). Type: Tanzania, Biharamulo/Mwanza District: Uzinza [Usinda], near French Mission, Usambiro, *Stuhlmann* 846, 849 (B†, syn.)

NOTE. *Gardner* in F.D. 2473 (Kenya, Machakos District: S slopes of Mbooni Ridge, Sept. 1930) and several other specimens have large ovoid acuminate buds up to 15 × 8 mm in addition to the normal fusiform ones – these are definitely galls and contain insect larvae – probably microlepidoptera.

17. **Ochna** sp. 17

Spindly shrub 2–2.7 m tall; flowering when quite or almost leafless; bark grey, ridged and lenticellate. Leaves elliptic-oblong to obovate-oblong, 2.4–4.5(–7.2) cm long, 1.2–2.5 cm wide, broadly rounded at the apex, narrowly rounded or ± cuneate at the base, rather obscurely distantly serrulate; lateral veins ± 25, ± prominent on both surfaces; tertiary venation reticulate, ± prominent; petioles 2.5–3.5 mm long; stipules narrowly triangular, 1.5 mm long, entire. Flowers 4–8 per fascicle, the fascicles on numerous short shoots themselves arranged on short branchlets, the whole forming a narrow panicle-like inflorescence along a main branch; pedicels about 1 cm long enlarging to 2 cm in fruit, jointed 1.5–3 mm from the base. Sepals yellow, oblong, 4–5 mm long, 2–3 mm wide, turning bright red and enlarging up to 9 mm long, 4 mm wide in fruit. Petals not seen. Anthers 1.7 mm long, shorter than the filaments, dehiscing by apical pores. Carpels 5; styles united save for 1 mm at apex. Drupelets subglobose to ellipsoid, 7.5 mm long, 6–6.5 mm wide.

KENYA. Lamu District: 88 km NE of Lamu, NW of Kiunga, 6 Aug. 1961, *Gillespie* 141! & Kiunga, Apr. 1931, *MacNaughton* 16 in F.D. 2567! & Ishakani ruins just N of Shakani, between Kiunga and Dar es Salaam, 5 Apr. 1980, *Gilbert & Kuchar* 5874!

DISTR. K 7 (Lamu); not known elsewhere

HAB. *Panicum pinifolium* dunes with *Salvadora* and *Zanthoxylum* thicket; scrub and savannah on sandy shore and above; 0–3 m

SYN. *O.* sp. A; Beentje, K.T.S.L.: 123 (1994)

NOTE. *MacNaughton* 16 has been determined by Robson as *O. ovata* F. Hoffm. but apart from the specialised habitat the flowers are more numerous per fascicle and the leaves more distinctly oblong. Only the three sheets cited have been seen, only one of which has a few leaves. More adequate material is needed to confirm that it is a distinct taxon.

18. **Ochna hackarsii** *Robyns & Lawalrée* in B.J.B.B. 18: 27 (1947); Robyns, F.P.N.A. 1: 614 (1948); van der Ben, Expl. Hydrobiol. Lacs Kivu, Edouard et Albert IV (1): 156 (1959); Bamps, F.C.B. Ochnaceae: 21 (1967). Type: Congo (Kinshasa), plain of Lake Edward, *Hackars* s.n. (BR, holo., K!, fragment)

Slender shrub 2–4.5 m tall; bark green, very smooth; slash pink, thin and fibrous; branches purplish brown, graceful, the youngest parts with dense small lenticels, later finely ridged. Leaves thin, copper-coloured when young, elliptic, 3–9 cm long, 1.5–4 cm wide, narrowed to an acute, subacute or ± rounded apex, rounded or

© The Board of Trustees of the Royal Botanic Gardens, Kew, 2005

broadly cuneate at the base, crinkly at margin when dry, crenate-serrate, the serrations often ending in inwardly curved teeth, lateral veins and reticulate tertiary venation not very evident, finely raised on both surfaces; petiole 1–3 mm long; stipules linear-lanceolate, 4–9 mm long, very shortly 2–3-fimbriate at apex, soon falling. Buds slender, fusiform, ± 7 mm long. Inflorescences 1(–2)-flowered; pedicel 1–1.5 cm (–2.2 cm in fruit) long, jointed 4–10 mm from base. Sepals elliptic, 5–6 mm long, 3–5 mm wide, becoming a brilliant dark rich red in fruit and up to 15 mm long, 8–9 mm wide. Petals yellow, obovate, 9 mm long, 6 mm wide. Anthers 1.5 mm long, opening by apical pores; filaments 2–2.5 mm long. Carpels 5; style 5–6 mm long, 5-fid at apex. Drupelets ovoid, 6–8 mm long, 5–6 mm wide.

UGANDA. Bunyoro District: Murchison Falls, Apr. 1933, *Eggeling* 1222 in F.D. 1328!; Ankole District: Kagera R., Kyanstone I., 26 Oct. 1963, *Langdale Brown* 161!; Mengo District: Singo, Bukomero, Apr. 1950, *Sangster* 1048!
DISTR. U 2, 4; E Congo (Kinshasa), Rwanda, Burundi
HAB. Fringing forest of *Piptadeniastrum africanum*, *Ficus*, *Albizia*, *Phoenix* etc.; 750–1300 m

SYN. *O. monantha* sensu I.T.U. ed. 2: 280 (1952), *non* Gilg

19. **Ochna leucophloeos** A. *Rich.* in Tent. Fl. Abyss. 1: 129 (1847), t. 29 (1851); Oliv. in F.T.A. 1: 318 (1868); F.P.S. 1: 187 (1950); Vollesen, Fl. Ethiopia & Eritrea 2 (2): 67, fig. 69. 2. 3 & 4 (1995). Type: Ethiopia, banks of R. Mareb, *Quartin Dillon* s.n. & banks of R. Taccazé, *Schimper* II. 1408 (both P, syn.)

Much branched shrub 3–3.6 m tall or tree to 8 m with grey or yellowish often powdery or scaly bark; branches grey-brown or yellowish, ± ridged and minutely flaky or ± powdery; youngest shoots purplish brown, lenticellate. Leaves drying pale brown or greenish brown, elliptic to slightly obovate, 4–7.5(–21) cm long, 2.3–4(–8) cm wide, subacute, obtusely acuminate or rounded at the apex and sometimes with short apiculum, narrowed to a truncate or slightly subcordate base or cuneate, margin serrate with incurved teeth; lateral veins ± 30, becoming strongly incurved at the margin, anastomosing or the marginal vein reaching near to the apex, together with the well marked subparallel tertiary venation very prominent on both surfaces; petiole 2–4 mm long, winged from decurrent leaf-base; stipules linear-triangular, 4–5 mm long, striate at apex, narrowly bifid, fairly soon deciduous. Flowers solitary or more often in 2–4(–6)-flowered subumbelliform inflorescences; pedicels slender, 1.5–4 cm long, jointed (5–)7–17 mm from the base. Sepals yellowish red, oblong to oblong-obovate or ovate, 4–6 mm long, 3 mm wide, 9–11 mm long, 4.5–6 mm wide in fruit, not spreading. Petals bright yellow, obovate, 5–7 mm long, up to 5 mm wide. Anthers 1.5–2 mm long, shorter than filaments, dehiscing by apical pores. Carpels 5; styles free at apex for just over 1 mm. Drupelets ellipsoid, 8–13 mm long.

SYN. *O. ardisioides* Webb, Frag. Fl. Aethiop.: 59 (1854); F.P.S. 1: 187 (1950). Type: Sudan, ? Fazogli, *Figari* s.n. (FI-W, holo.)
    *O. micropetala* Martelli, Fl. Bogos: 13 (1886). Type: Ethiopia, Djeladjeranne, *Schimper* III, 1738 (FI-W, holo., FT, K!, P, iso.)

NOTE. I have not seen authentic material of Webb's species but both Oliver and Vollesen include it under *O. leucophloeos*.

subsp. **ugandensis** Verdc. **subsp. nov.** a subspecie typica foliis ellipticis 4–7.5 cm longis, 2.3–4 cm latis basi rotundatis vel leviter subcordatis differt. Typus: Uganda, Acholi District: Adilang, *Greenway & Hummel* 7338 (K!, holo., EA, iso.)

Shrub to 3.6 m. Leaves elliptic, 4–7.5 cm long, 2.3–4 cm wide, rounded to slightly subcordate at the base.

UGANDA. Acholi District: Mt Madi, Mar. 1935, *Eggeling* 1740 in F.D. 1615! & Adilang, 11 Apr. 1945, *Greenway & Hummel* 7338!

© The Board of Trustees of the Royal Botanic Gardens, Kew, 2005

Distr. **U** 1; not known elsewhere

Hab. Open woodland of *Xerophyta, Tarenna, Sterculia, Psychotria* and *Terminalia* on rocky hillsides; ± 900 m

Syn. *O. ovata* sensu I.T.U. ed. 2: 280 (1952) pro parte, *non* F. Hoffm.

Note. Robson thought that the Uganda material was a sp. aff. *O. leucophloeos* and both sheets have been annotated as that species by Wadhwa. Further material is needed to assess its status.

20. **Ochna pseudoprocera** *Sleumer* in F.R. 39: 277 (1936); T.T.C.L.: 383 (1949); Robson in Bol. Soc. Brot. sér. 2, 36: 21 adnot. (1962) and in F.Z. 2: 239 adnot. (1963); Vollesen in Opera Bot. 59: 25 (1980). Type: Tanzania, Lindi District: 60 km W of Lindi, Rondo [Muera] Plateau, *Schlieben* 6178 (B†, holo.)

Shrub or tree 1–5(–10) m tall; branches slender, grey, strongly lenticellate. Leaves thin, lanceolate or oblong-elliptic or rarely subovate-lanceolate, 3.5–6.5 cm long, 1–2.4 cm wide, acutely acuminate at the apex, cuneate at the base, rather distantly serrate, the teeth rather obscure, appearing shallowly crenate to the naked eye; lateral veins ± 30, together with reticulate tertiary venation finely raised on both surfaces; petiole slender, 3–4 mm long; stipules linear, 4 mm long, bifid. Flowers 2–3 at ends of short lateral branchlets; pedicels 1–1.5 cm long, jointed about 1 mm from base. Sepals red in fruit, unequal, narrowly oblong-elliptic to broadly elliptic, 1.2–1.5 cm long, 0.4–1.1 cm wide. Petals yellow, narrowly obovate, 7 mm long, 3.5 mm wide. Anthers 1–1.5 mm long, dehiscing by apical pores (see note); filaments 2.5–3.5 mm long. Styles united save for apical free branches ± 0.5 mm long. Drupelets black, ellipsoid, 8–9 mm long, 6–7 mm wide.

Tanzania. Kilwa District: Nakilala Thicket, 14 Dec. 1975, *Vollesen* MRC 3075! & Malemba Thicket, 24 Jan. 1977, *Vollesen* MRC 4376!; Lindi District: Sudi, May 1943, *Gillman* 1464!
Distr. **T** 8; not known elsewhere
Hab. Deciduous coastal thicket on sand, woodland; 300–400 m

Note. Robson mentions this is related to a species with porate anthers and MRC 3075 certainly has porate anthers but Sleumer clearly states anthers dehiscing by longitudinal slits. Vollesen's and my interpretation of the species rests on Robson's identification of *Gillman* 1464.

21. **Ochna rovumensis** *Gilg* in E.J. 33: 246 (1903); T.T.C.L.: 384 (1949); Robson in F.Z. 2: 238, t. 44 fig. B (1963); Vollesen in Opera Bot. 59: 25 (1980). Type: Tanzania, Masasi District: Rovuma R., near Makotschera, *Busse* 1298 (B†, holo., EA!, iso.)

Shrub or tree 1.5–15 m tall with smooth grey bark with lighter patches; branches ± spreading or ascending, whitish or pale brown, glabrous, slightly peeling or puberulous in some specimens from Flora Zambesiaca area. Leaves thin, elliptic or oblong-elliptic, 3–7 cm long, 1.3–3.5 cm wide, rounded to acute at the apex and sometimes shortly mucronate-aristate at tip, cuneate to rounded at the base, subentire or finely serrulate at margin, main and subsidiary lateral veins and intervening tertiary reticulate venation ± prominent on both surfaces; petiole 2–4 mm long. Flowers solitary on very short axillary shoots near tops of branches; pedicels 1–1.8 cm long, glabrous or rarely puberulous (not in East Africa), jointed at base. Sepals broadly elliptic, rounded, (0.9 fide Robson–)1.5–2 cm long, 1.2 cm wide, navicular; remaining ± closed so flower is ellipsoid, eventually becoming red and 2–3 cm long, 1.5 cm wide, at length ± spreading. Petals bright yellow, obovate, 26 mm long, 15 mm wide. Anthers 1.5–2 mm long, $^1/_4$–$^1/_2$ as long as filaments, dehiscing by apical pores. Carpels 5; styles free towards apex, the ends spreading, with small stigmas. Drupelets ovoid-cylindric, 10 mm long, 6 mm wide.

© The Board of Trustees of the Royal Botanic Gardens, Kew, 2005

TANZANIA. Kilosa District: along R. Ruaha, 10 Jan. 1956, *Benedicto* 101!; Kilwa District: ± 19 km SSW of Kingupira, 28 Nov. 1976, *Vollesen* MRC 4170!; Masasi District: near Makotschera, Feb. 1901, *Busse* 1298!
DISTR. **T** 6, 8; Mozambique, Malawi, Zimbabwe
HAB. Rocky riverine thicket; 150–600 m

22. **Ochna holstii** *Engl.* in Abh. Preuss. Akad. Wiss: 69 (1894) & in P.O.A. C: 273 (1895); Gilg in E.J. 33: 234, 241 (1903); T.T.C.L.: 382 (1949); I.T.U. ed. 2: 279 (1952); K.T.S.: 337 (1961); Robson in F.Z. 2: 240 (1963); Bamps, F.C.B. Ochnaceae: 8 (1967); du Toit & Obermayer, F.S.A. 22: 5 (1976); Blundell, Wild Fl. E Afr.: 62, fig. 276 (1987); K.T.S.L.: 121 (1994); Vollesen in Fl. Ethiopia & Eritrea 2 (2): 66, fig. 69.1.4 (1995). Type: Tanzania, Lushoto District: West Usambaras, Mbalu, *Holst* 2601 (B, holo., K!, iso.)

Tree or shrub 3–27 m tall with smooth grey or greybrown bark; branches purplish brown, somewhat angular, usually glabrous or rarely puberulous when young; bark not peeling, lenticellate; slash pink to dull red. Leaves mostly rather thin, oblanceolate, obovate or elliptic to oblong, (1.5–)5–13 cm long, (0.7–)1.2–4.3 cm wide, usually acute to narrowly acutely acuminate, but can be more obtuse in small-leaved forms in south, cuneate to ± rounded at the base, margin densely curved-serrulate; lateral veins 20–25, usually ± at right angles to mid-rib together with the dense tertiary venation prominent above; petiole 1–3(–3.5) mm long, ± slender and often appearing longer since leaves can be borne on very short slender spur shoots, but with a small bud at junction with petiole. Flowers 5–20 in racemiform inflorescences with rhachis up to 2 cm long, less often subumbellate; pedicels 1.3–4 cm long, articulated at base (first flower) or within 3 mm of it, glabrous or rarely puberulous. Sepals oblong-elliptic, unequal, 6–9 mm long, 3–5 mm wide, becoming deep red in fruit, 9–18 mm long, (3.5–)6–8 mm wide. Petals pale to bright yellow, obovate, 8–12 mm long, 3–6 mm wide. Anthers 1–2 mm long, ± ¹/₂ length of the filaments, dehiscing by longitudinal slits. Carpels 5(–6) with styles usually completely united; stigma subglobose or 5–6-lobed. Drupelets black, ellipsoid-cylindric, less often ovoid or subglobose, 8–14 mm long, 5–9 mm wide, inserted at or near base.

UGANDA. Acholi District: SE Imatong Mts, Aringa R. headwaters, 6 Apr. 1945, *Greenway & Hummel* 7296! & Imatong Mts, Apr. 1938, *Eggeling* 3539!; Mbale District: Suam Valley, May 1933, *Dale* in F.D. 3121!
KENYA. Elgeyo/Uasin Gishu Districts: Kaptagat, S Elgeyo Forest, June 1934, *McIntyre* in F.D. 3289!; Kiambu District: top of Kedong Rift wall about 37 km from Nairobi, 9 Feb. 1963, *Verdcourt* 3582!; Kisumu-Londiani District: Tinderet Forest Reserve, 15 June 1949, *Maas Geesteranus* 4971!
TANZANIA. Lushoto District: East Usambaras, Amani, 19 Jan. 1942, *Greenway* 6432!; Kondoa District: Kinyassi Mt, 2 Jan. 1928, *B.D. Burtt* 1806!; Njombe District: Njombe, 18 Dec. 1931, *Lynes* 115!
DISTR. **U** 1, 3; **K** 1, 3–7; **T** 2–8; Congo (Kinshasa), Rwanda, Burundi, Sudan, Ethiopia, Malawi, Zambia, Zimbabwe, Mozambique and South Africa
HAB. Various types of upland evergreen forest, drier forest, montane forest, rain-forest; also atypically a small shrub in upland grassland and thicket, on granite rocks by waterfalls and in Songea a form in crevices of rock outcrops; (250– (see note)) 900–2350 m

SYN. *O. prunifolia* Engl., P.O.A. C: 273 (1895); Gilg in E.J. 236 (1903); T.T.C.L.: 384 (1949). Types: Kenya, Kitui District: Kitui, *Hildebrandt* 2825 (B†, syn.); Tanzania, District unclear, Rombo Mku, *Volkens* 1960 (B†, syn.) & District unclear, Mkulia, *Volkens* s.n. (B†, syn.)
　*O. longipes* Bak. in K.B. 1897: 247 (1897); Brenan in Mem. N.Y. Bot. Gard. 8: 234 (1953); White, F.F.N.R.: 251 (1962). Type: Malawi, Mt Malosa near Zomba, *White* 429 (K!, holo.)
　*O. shirensis* Bak. in K.B. 1897: 247 (1897). Type: Malawi, Mt Zomba & Mt Malosa, *White* 430 (K, holo.!)

© The Board of Trustees of the Royal Botanic Gardens, Kew, 2005

*O. acutifolia* Engl. in E.J. 28: 433 (1900). Type: Tanzania, Morogoro District: Uluguru, near Nghweme, *Stuhlmann* 8852 (B†, syn.), & Lushoto District: West Usambaras, *Buchwald* s.n. (B†, syn., BM, K!, isosyn.)*

*O. densicoma* Engl. & Gilg in E.J. 33: 241 (1903); T.T.C.L.: 382 (1949). Type: Tanzania, Lushoto District: East Usambaras, Nderema, *Scheffler* 169 (B†, holo., BM!, EA!, K!, iso.)

*O. keniensis* Sleumer in N.B.G.B. 12: 69 (1934); K.T.S.: 338 (1961); K.T.S.L.: 121 (1994). Type: Kenya, Mt Kenya S slope, near Rupingazi R., *R.E. & T.C.E. Fries* 2031 (UPS!, holo.)

NOTE. As Robson has pointed out *O. holstii* is very variable. The specimens with large acutely acuminate leaves which are mostly tall forest trees present no difficulty but in many often drier areas small-leaved specimens which are often no more than small shrubs are so different that it is difficult to believe they are the same species; attempts at sorting out varieties have failed. *Ruffo & Kisena* 2778 from Muva Mbizi Forest Reserve at 1880 m collected on 21 Nov. 1967 is a shrub only 2 m tall with small elliptic obtuse leaves up to 3.8 × 1.5 m. A sterile specimen *Gilchrist* in F.H. 1734 from the Calderara [sic] clearing near Mufindi has large ovoid inflorescence buds 9 × 5 mm very different from the narrower ones of the northern forest trees but Robson has accepted it as *O. holstii*. *Vollesen* MRC 4410 from Mayombera Forest Reserve, lowland forest of *Erythrophleum suaveolens, Isoberlinia, Treculia* and *Parkia* is from 250 m, the lowest altitude noted, and has blunt leaves. *Cribb et al.* 11157 (Tanzania, **T** 6, Ulanga District: Sali, N side of E peak of Ngongo Mt at 1350 m, floor of montane forest, 23 Jan. 1979) has been referred to *O. holstii* but the young stems and 25 mm long pedicels are distinctly pubescent and it is a shrub to 1.5 m; the ellipsoid-cylindric drupelets have attachments to distinctly one side of the base. Anthers not known but presumably opening by splits. Much further fieldwork needs doing on these shrubby variants.

23. **Ochna oxyphylla** *Robson* in Bol. Soc. Brot., Sér. 2, 36: 21 (1962) & in F.Z. 2: 241 (1963). Type: Tanzania, Morogoro District: Uluguru Mts, Bondwa Hill, *Drummond & Hemsley* 1764 (K!, holo., K!, SRGH, iso.)

Shrub or small tree (0.4–)2–8 m tall with grey to reddish brown rough or slightly fissured bark; branches reddish-brown, slender, ribbed, puberulous or rarely glabrous, later purplish with numerous pale lenticels. Leaves fairly thin, drying green or brown or sometimes blue-green, narrowly elliptic to oblanceolate, 2–6.7 cm long, 1–2 cm wide, acute to very acutely acuminate at the apex, rounded to broadly cuneate at the base, the margin densely spinulose-serrate with curved teeth; lateral veins ± 25, together with the dense reticulate tertiary venation prominent above but hardly beneath; petiole ± obsolete or up to 1.5 mm long; stipules orange-brown, narrowly oblong-lanceolate, 9 mm long, 2.5 mm wide, soon deciduous. Flowers 1–5 in short raceme-like or subumbellate inflorescences with rhachis 0–2 mm long; pedicels 1.1–2.4 cm long, jointed at base or up to 2 mm from base, puberulous or glabrous. Sepals oblong-elliptic, 5–7 mm long, 4 mm wide, becoming pink or red in fruit and up to 9–11 mm long, 5–6 mm wide. Petals yellow, narrowly obovate, 7–10 mm long, 3–4.5 mm wide. Anthers 1.3–2 mm long, slightly longer or shorter than the filaments, dehiscing by longitudinal slits. Carpels 5, the styles completely united; stigma subglobose or 5-lobed. Drupelets black, ellipsoid, 6–7 mm long, 4 mm wide.

TANZANIA. Morogoro District: Uluguru Mts, above Bunduki, Mgeta R., 1 Jan. 1975, *Polhill & Wingfield* 4634!; Mbeya District: Poroto Mts, Ikuyu, about 8 km N of Irambo, 10 Feb. 1979, *Cribb et al.* 11380!; Njombe District: Livingstone Mts, 1.5 km S of Msalaba Mt, above Luana, 23 Nov. 1992, *Gereau et al.* 5126!
DISTR. **T** 6, 7; N Malawi
HAB. Upland rain-forest, riverine forest, grassland/montane forest boundaries, *Parinari* woodland; 1650–2400 m

NOTE. *Ede* 42 (Tanzania, Iringa District, Nyumbanyitu, 22 Sept. 1958) cited by Robson in his original description is described as a tree to 70 feet and 6 feet girth but I think there must be some confusion with the field note.

* Although the *Buchwald* specimen was mentioned in a note I think it has to be looked on as a syntype. BM & K sheets are numbered 350.

© The Board of Trustees of the Royal Botanic Gardens, Kew, 2005

24. **Ochna stolzii** *Engl.*, V.E. 3 (2): 480 (1921) (in clav.); Gilg in E & P. Pf. ed. 2, 21: 68 (1925); Sleumer in N.B.G.B. 12: 69 (1934); Robson in F.Z. 2: 241 (1963). Type: Tanzania, Rungwe District: Kyimbila, Tandala, *Stolz* 2212 (B†, holo.)

Shrub or small shrublet 0.35–2 m tall with smooth grey-brown bark and ascending reddish brown, puberulous branches, later greyish brown, glabrous and lenticellate. Leaves often drying metallic bluish green, somewhat coriaceous, small, elliptic to obovate, 1.2–5.5 cm long, 0.5–2.3 cm wide, acute to rounded at the apex, broadly cuneate to rounded at the base, margin spinulose-serrulate; lateral veins 13–15, ± prominent on both surfaces, the reticulate tertiary venation more obscure; petiole ± obsolete or up to 1.5 mm long; stipules linear, 4.5 mm long, deciduous. Flowers solitary or up to 3 in fascicles or rarely up to 7 in racemiform inflorescences with rhachis up to 8 mm long; pedicels 0.8–2.5 cm long, jointed near the base, glabrous or puberulous, often reflexed in fruit. Sepals oblong-elliptic, 6 mm long, 3–4 mm wide, becoming carmine red in fruit and 8–12 mm long, 5–6 mm wide. Petals yellow, obovate, 8–12 mm long, 3.5–8 mm wide. Anthers ± 1.5 mm long, slightly shorter than the filaments, dehiscing by longitudinal slits. Carpels 5; styles completely united; stigma subglobose. Drupelets black, elliptic, 7–8 mm long, 4–5 mm wide.

TANZANIA. Iringa District: Mufindi, Lake Ngwazi, 28 Mar. 1991, *Bidgood & Vollesen* 2157!; Njombe District: Njombe, 7 Dec. 1931, *Lynes* D81! & Njombe–Milo road, 28 Jan. 1961, *Richards* 14019!
DISTR. **T** 7; N Malawi
HAB. Hillside grassland, grassland with dry temperate rain-forest remnants, *Brachystegia* woodland, steep banks; 1800–1950(–2100) m

NOTE. The largest leaves I have seen are 4.3 cm long, 1.9 cm wide; the upper measurements are those of Sleumer. The highest altitude is that given by Stolz. Sleumer does not mention pubescence on the young branches and pedicels and some doubt must remain about Robson's interpretation. *Gereau & Kayombo* 4021 (Tanzania, Njombe District: Livingstone Mts, Ligala Mt, 13 Feb. 1991, dense woodland at 2000 m) has leaves up to 4.5 × 2 cm and the pedicels and young branches are puberulous but the inflorescences have elongated axes and are arranged on closely placed nodules below the leaves forming a compound inflorescence ± 8 × 7 cm; the pedicel joint is 7 mm above the base and I have referred it to a small-leaved *O. holstii*. Sleumer does not give the length of the joint for *O. stolzii*. However, Robson has annotated some small-leaved specimens with basal joints as *O. holstii*. The true nature of these atypical southern *O. holstii* needs much further fieldwork.

25. **Ochna polyneura** *Gilg* in E.J. 33: 240 (1903); T.T.C.L.: 383 (1949); Robson in F.Z. 2: 242 (1963); Vollesen in Opera Bot. 59: 25 (1980). Types: Tanzania, (District not clear) Mbarangandu region, *Busse* 681 (B†, syn., EA!, isosyn., K, photo!)

Small tree, shrub or subshrub 0.3–6(–8) m tall with pale grey rough deeply fissured bark; branches reddish to purple-brown, mostly smooth, becoming whitish, with epidermis exfoliating, usually lenticellate, sometimes minutely papillose-puberulous. Leaves often bronze-tinged and sometimes drying quite dark bluish green, thin, obovate to oblanceolate or elliptic, (2.5–)3.5–10(–12) cm long, 1–3(–5) cm wide, narrowly rounded at the apex or subacute, cuneate to narrowly rounded at the base, margin densely usually rather shallowly serrulate; lateral veins ± 30, ± at right angles to midrib then more strongly curved upwards; tertiary venation closely reticulate, prominent above, much less so beneath; petiole 1–2 mm long. Buds large, ovoid, 15 mm long, 8 mm wide, acute, imbricate, the bracts strongly striate. Flowers (5–)6–10(–14) in elongated raceme-like inflorescences with rhachis 0.3–2.5(–3.5) cm long, rarely subumbellate; pedicels slender (1–)1.8–2.8(–3.5) cm long, papillose-puberulous or glabrous, jointed 1–7 mm from the base. Sepals pale green, oblong-elliptic, 5–7 mm long 1.5–2 mm wide, rounded at the apex becoming deep red in fruit and enlarging to 10–12(–15) mm long, 7 mm wide. Petals pale yellow, obovate to round, 9–13 mm long, 5–8 mm wide. Anthers 1–2 mm long, about $^{1}/_{2}$ the length of the filaments,

© The Board of Trustees of the Royal Botanic Gardens, Kew, 2005

dehiscing by longitudinal slits. Carpels 5(–6–7); styles completely united; stigma subglobose. Drupelets black, ovoid or somewhat curved, (7–)8.5–10 mm long, 5–6 mm wide. Fig. 3 (p. 27).

TANZANIA. Morogoro District: near Kingolwira Station, 17 Nov. 1956, *Welch* 336!; Uzaramo District: Lulenzi, 45 km WNW of Dar es Salaam, 3 Oct. 1977, *Abdallah* 208!; Songea District: ± 12 km E of Songea, Nonganonga stream, 27 Dec. 1955, *Milne-Redhead & Taylor* 7915! & 21 Jan. 1956, 7915A!
DISTR. **T** 6–8; Malawi, Zimbabwe, Mozambique
HAB. Wooded grassland with *Isoberlinia*, *Combretum* and *Acacia nigrescens*, mixed woodland of *Brachystegia*, *Pterocarpus*, *Burkea*, *Terminalia*, *Pteleopis* etc., *Parinari* woodland, also deciduous coastal thicket on sand; (80–)130–1050(–2000) m

SYN. *O. hylophila* Gilg in E.J. 33: 242 (1903); T.T.C.L.: 383 (1949). Type: Tanzania, Songea District: Mampyui road on E Ungoni border, *Busse* 732 (B†, holo., EA!, iso., K, photo.!)

26. **Ochna puberula** *Robson* in Bol. Soc. Brot. sér. 2, 36: 25 (1962) & in F.Z. 2: 244, frontispiece (1963). Type: Zambia, Mbala [Abercorn], Kawimbe, *Richards* 10235 (K!, holo.)

Shrub or small tree 0.5–7.5 m tall but possibly much taller (see note) with grey smooth or ± reticulately fissured bark; branches reddish brown, somewhat 4-angled and ± densely papillose-puberulous at first, becoming glabrous, striate or shallowly fissured, densely lenticellate. Leaves fairly thin, often slightly glaucous, usually drying dark bluish green, obovate to oblanceolate, elliptic or oblong-elliptic, (2.2–)3.6–7.5 cm long, 1–2.7(–2.9) cm wide, narrowed or slightly acuminate to a rounded apex or subacute, cuneate at base or narrowed to a rounded or truncate base, densely curved-serrulate at margin; lateral veins ± 25 with numerous intermediaries, together with closely reticulate tertiary venation, prominent on both surfaces; petiole 0.5–2 mm long. Flowers sometimes (perhaps always?) precocious, 2–8 in pseudo-umbellate or raceme-like inflorescences; pedicels 1–2.7(–3.5) cm long, jointed at base or up to 3 mm from base, sometimes with a deciduous stipule-like bract at the junction, papillose-puberulous, particularly the jointed portion. Sepals broadly elliptic, 3–7 mm long, 3.5–6 mm wide, rounded, becoming yellow then orange-red to crimson in fruit, 10–14 mm long, 8–11 mm wide, remaining tightly imbricated around the developing drupelets forming an ellipsoid structure but eventually ± spreading. Petals bright yellow, obovate, (5–)7–13 mm long, 3–7.5 mm wide. Anthers sometimes drying dark bluish green, 1–1.5 mm long, shorter than the filaments, dehiscing by longitudinal slits. Carpels 5–7; styles completely united or rarely free at apex; stigma subglobose or 5–7-lobed. Drupelets black, ellipsoid, 8–10 mm long, 5–6 mm wide.

TANZANIA. Ufipa District: Muva-Mbiza Forest Reserve, 21 Nov. 1987, *Ruffo & Kisena* 2780!; Kondoa District: Simbo Hills, 8 Jan. 1928, *B.D. Burtt* 1027!; Njombe District: Makete, Lower Ndumbi Valley, 14 Dec. 1986, *Lovett & Congdon* 1109!
DISTR. **T** 1–7 (2, 3, 6 fide EA); Zambia and Zimbabwe
HAB. *Brachystegia – Pterocarpus* woodland particularly on exposed granitic ridges in rock crevices etc.; "high forest" fide *Burtt* 921 (see note); (1050–)1400–2100 m

SYN. *O. longipes* sensu Norlindh in Bot. Notis. 1948: 32 (1948), *non* Bak.
    *O. sp. nr. holstii*; Brenan, T.T.C.L.: 382 adnot. (1949)
    *O.* sp.; White, F.F.N.R.: 252 (1962) pro parte excl. *White* 2806

NOTE. *Congdon* 303 (Tanzania, Mbeya District: Madibira Hills, 9 Dec. 1990, *Brachystegia* woodland at 1400 m) flowering when leafless or leaves very juvenile appears to be this species. Robson has annotated *B.D. Burtt* 921 (Tanzania, Kondoa District: Kinyassi Mt, 2 Jan. 1928) as *O. puberula* and it looks identical but Burtt's field note states 'tall tree 50–80 ft tall in high forest' and I think some confusion may have occurred; of 1053 (Kondoa District: near Mnenya [Mnenia]) on scarp, 13 Jan. 1928) he states 'spectacular shrub coppicing from a large stump'; this must have been a large tree at some time in the past and is possibly the same as 921.

© The Board of Trustees of the Royal Botanic Gardens, Kew, 2005

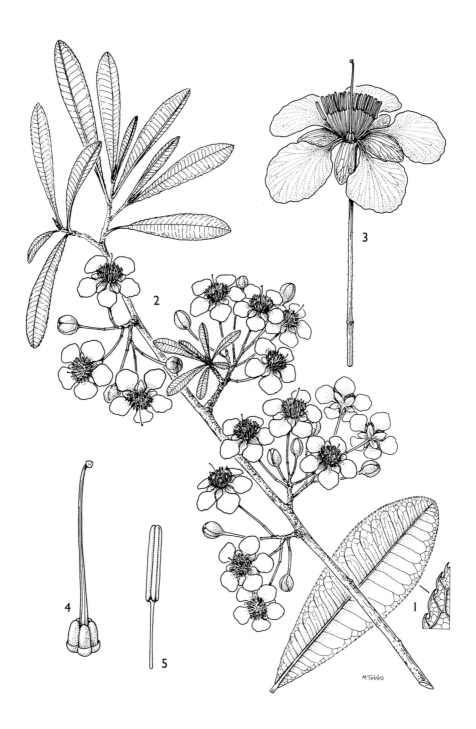

FIG. 3. *OCHNA POLYNEURA* — **1**, mature leaf, × ²/₃; **2**, flowering branch, × ²/₃; **3**, flower, × 2; **4**, style and ovary, × 8; **5**, stamen, × 10. 1 from *Milne-Redhead & Taylor* 7915a; 2–5 from *Milne-Redhead & Taylor* 7915. Drawn by Margaret Tebbs.

© The Board of Trustees of the Royal Botanic Gardens, Kew, 2005

27. **Ochna afzelioides** *Robson* in Bol. Soc. Brot. sér. 2, 36: 23 (1962) & in F.Z. 2: 242 (1963). Type: Tanzania, Buha District: Kasulu, *Rounce* B 3 (K!, holo., EA, iso.)

Shrub or tree up to 6 m tall with smooth grey bark; branches purplish- or yellow-brown, lenticellate, papillose-puberulous or glabrous but later quite glabrous. Leaves thin, oblanceolate or elliptic or obovate-oblanceolate, 4.5–9 cm long, 1.8–3.3 cm wide, acute or acuminate at the apex, cuneate to rounded at the base, margin densely spinulose-serrate or curved-serrulate; lateral veins and densely reticulate tertiary venation finely prominent above and below but not very conspicuous; petiole 1.5–2 mm long. Flowers (5–)7–10 in a racemiform inflorescence with rhachis 9–20 mm long; pedicels 1.4–2.6 cm long, jointed about 1–5 mm from the base, papillose-pubescent or glabrous. Sepals drying blackish, ovate or elliptic, 5–7 mm long, 2.5–3.5 mm wide, acute or rounded, red in fruit and enlarging to 7–9 mm long, 3–5 mm wide. Petals pale yellow, obovate, 8–10 mm long, 5–6 mm wide. Anthers 1.5 mm long, about $\frac{1}{2}$ as long as filaments, dehiscing by longitudinal slits. Carpels 5–8; styles completely united; stigma depressed subglobose. Drupelets ellipsoid, 6–8 mm long, 4–5 mm wide, inserted below the centre.

TANZANIA. Bukoba District: Kiamawa, Sept.–Oct. 1935, *Gillman* 458!; Buha District: Kasulu, Oct. 1930, *Rounce* B 3!; Kigoma District: near Tubila railway station, Nov. 1956, *Procter* 588!
DISTR. **T** 1, 4; N & W Zambia (see note)
HAB. *Brachystegia* woodland and evergreen forest edges; ± 1200 m

NOTE. Robson recorded this species from Congo (Kinshasa), Shaba [Katanga] but Bamps (F.C.B., Ochnaceae: 8 (1967)) refers the specimens concerned to *O. puberula* Robson. All the Tanzanian material seen has been completely glabrous. Two further specimens which I think may belong to *O. afzelioides* extend the distribution to **K** 2 and **T** 7 – *Newbould* 7172, Kenya, Turkana District, Muruanysigar, 16 Feb. 1965 at 2100 m and *D.W. Thomas* 3922, Tanzania, Iringa District, Mwanihana Forest Reserve, above Sanje, 10 Oct. 1984 at 1400–1700 m. Both these have very acutely acuminate leaves.

28. **Ochna leptoclada** *Oliv.* in F.T.A. 1: 318 (1868); Gilg in E.J. 33: 233 (1903); Brenan, T.T.C.L.: 383 (1949) & in Mem. N.Y. Bot. Gard. 8: 234 (1953); White, F.F.N.R.: 251, fig. 43f (1962); Robson in F.Z. 2: 243 (1963); Bamps in F.A.C., Ochnaceae: 12 (1967); Vollesen in Opera Bot. 59: 25 (1980). Types: Malawi, Manganja Hills, Margomero, *Meller* s.n. & Malabve [Maravi] country, *Kirk* s.n. (both K!, syn., EA, photo.)

Shrub or rhizomatous shrublet 0.3–1.3 m tall with brown bark; shoots often caespitose, branched, greyish white with papery epidermis often peeling. Leaves subcoriaceous, mostly ± glaucous, obovate to obovate-oblanceolate, less often oblong, (3–)4–12(–14) cm long, 1.2–4.5(–5.2) cm wide, rounded or subacute at the apex, rarely ± emarginate and minutely apiculate, cuneate and somewhat twisted or curved at base, entire or remotely spinulose serrate; lateral veins 20–30, together with very finely reticulate tertiary venation prominent on both surfaces; petiole 2–4 mm long, ± thickened and grooved beneath; stipules triangular, 1–2 mm long. Flowers sometimes appearing when almost leafless, 1–3(–4) in false umbels which often form panicles; pedicels 0.9–2.8 cm long, jointed at the base. Sepals oblong-elliptic, 3–5 mm long, 2–2.5 mm wide, rounded at apex, turning bright deep crimson in fruit and enlarging to 9–15 mm long, 6 mm wide. Petals bright yellow, elliptic to obovate, 6–9 mm long, 4–6 mm wide. Anthers 1–2 mm long, $\frac{2}{3}$ as long as filaments, dehiscing by longitudinal slits. Carpels 5; styles completely united; stigma globose. Drupelets black, subglobose, 6–8 mm long, 5–6 mm wide, inserted at base, wrinkled when dry.

UGANDA. West Nile District: Aringa, Mt Kee CFR, 20 Feb. 1955, *Dale* 867 (fide EA)
TANZANIA. Mwanza District: Geita, Butundwe, Oct. 1949, *Watkins* 316!; Kigoma District: S end of Kigoma Bay, 14 June 1980, *Hooper et al.* 1975!; Ulanga District: Msolwa Camp, 2 Nov. 1977, *Vollesen* 4754!

© The Board of Trustees of the Royal Botanic Gardens, Kew, 2005

DISTR. **U** 1; **T** 1, 4, 6–8; E Congo (Kinshasa), Rwanda, Burundi, Sudan, Zambia, Malawi, Mozambique, Zimbabwe

HAB. *Brachystegia, Isoberlinia, Uapaca, Pterocarpus* etc. woodland, *Protea – Faurea* wooded grassland, sometimes on eroded ironstone soils, old cultivations; 250–1650 m

SYN. *O. fruticulosa* Gilg in E.J. 33: 238 (1903); T.T.C.L.: 382 (1949). Type: Tanzania, Biharomulo District: Ukome I., SW Creek, Niansa, *Stuhlmann* 882 (B†, holo.)

NOTE. No authentic material of *O. fruticulosa* exists but I am certain the above synonymy is correct. Topotypic material is needed to prove it beyond all doubt. *Richards & Arasululu* 26148 (Tanzania, Mpanda District: Uruwira–Mpanda, km 27) has much more prominently serrate leaf margins but I am sure it belongs here; a suggestion it was *O. cyanophylla* is not correct.

29. **Ochna cyanophylla** *Robson* in Bol. Soc. Brot., sér. 2, 36: 27 (1962) & in F.Z. 2: 244 (1963). Type: Zimbabwe, N Mazoe, *Stables* 14/56 (K!, holo.; SRGH, iso.)

Small tree 3.5–6 m tall with rather rough pale brown bark; branches yellow-brown, striate or ± ribbed, lenticellate but later surface peeling in thin papery strips. Leaves drying dark bluish green, ± subcoriaceous, obovate or oblanceolate or oblong-elliptic, 5.5–12 cm long, 2.3–5 cm wide, narrowed to a rounded apex, attenuate-cuneate at the base; margin crenulate-serrate; lateral veins close and prominent; tertiary venation densely reticulate, rather obscure to prominent; petiole 1–2 mm long; stipules narrowly triangular, folded, 4 mm long, entire, soon deciduous. Flowers (1–)3–7 in very short racemiform or subumbellate inflorescences with rhachis up to 5 mm long; pedicels 1.3–2(–3) cm long, jointed at or very near base. Sepals oblong-elliptic, 4–6 mm long, 2–4 mm wide, rounded, becoming deep cherry red in fruit, 8–13 mm long, 5–8 wide. Petals deep yellow, obovate, 7–11 mm long, 5–7 mm wide. Anthers 1–1.5 mm long, $\frac{1}{3}$–$\frac{1}{2}$ as long as the filaments, dehiscing by longitudinal slits. Carpels 6–8 with styles completely united; stigma subglobose. Drupelets ellipsoid, 9–10(–13) mm long, 6.5–8 mm wide.

TANZANIA. Mpwapwa District: Mpwapwa, 27 Feb. 1933, *Mr & Mrs Hornby* 499!; Iringa District: Iringa, 26 Jun. 1932, *Lynes* I.h. 14! & Magangwe*, 10 Dec. 1970, *Greenway & Kanuri* 14752!
DISTR. **T** 5, 7; Zimbabwe
HAB. *Brachystegia – Isoberlinia* and *Combretum – Terminalia – Strychnos* woodland; 1200–1500 m

30. **Ochna katangensis** *De Wild.* in Ann. Mus. Congo, Bot. Sér. 4, 1: 89, t. 33, figs 5–6 (1903); Gilg in E.J. 33: 232 (1903); Robson in F.Z. 2: 245 (1963); Bamps, F.C.B. Ochnaceae: 10 (1967). Type: Congo (Kinshasa), Shaba [Katanga], plains of the high plateau, *Verdick* s.n. (BR, holo.)

Pyrophytic subshrub 15–25(–45) cm tall but sometimes flowering at ground level, sometimes forming cushions from a many-headed woody rootstock; bark grey-brown, ± smooth or rough. Leaves narrowly elliptic, narrowly oblanceolate or narrowly oblong, (3–)4–12 cm long, 0.9–2.8 cm wide, narrowly rounded to acute at the apex, narrowly cuneate at the base, serrate with close curved teeth; lateral veins 12–20, prominent on both surfaces with very close reticulate or partly subparallel tertiary venation between, sometimes reddish; petiole 2 mm long; stipules linear-triangular, 3–7 mm long, entire or ± divided, deciduous. Flowers solitary or 2–5 in umbellate or shortly racemose inflorescences axillary on second-year growths; pedicels 1.5–4 cm long, articulated at or near base. Sepals elliptic or ± round, 5–8(–10) mm long, 2–3 mm wide, becoming red and up to 15 × 8 mm in fruit. Petals yellow or orange-yellow, obovate, (8–)12–13(–20) mm long, (4–)7–8(–13) mm wide. Anthers straight, 1.8–3 mm long, $\frac{1}{3}$–$\frac{2}{3}$ as long as the filaments, dehiscing by longitudinal slits. Carpels 5; styles

* Given as Mbeya District in F.T.E.A. Gazetteer.

© The Board of Trustees of the Royal Botanic Gardens, Kew, 2005

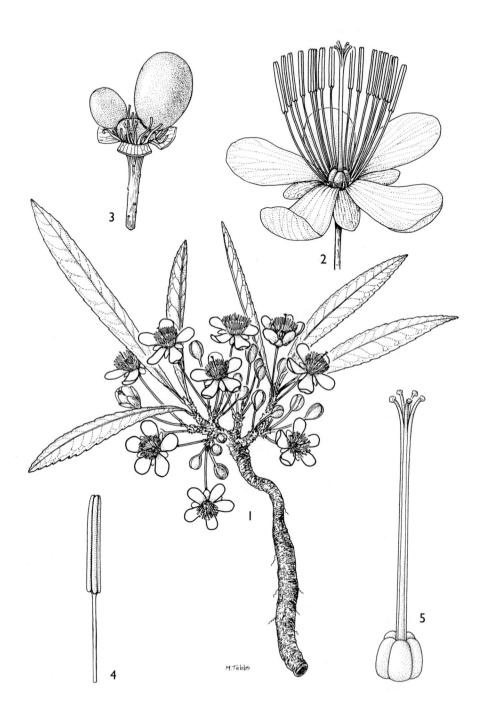

FIG. 4. *OCHNA KATANGENSIS* — **1**, habit, × ²/₃; **2**, flower, × 3; **3**, carpels, × 3; **4**, stamen, × 8; **5**, ovary and styles, × 8. 1, 2, 4, 5 from *Bullock* 3424; 3 from *Brummitt* 12007. Drawn by Margaret Tebbs.

© The Board of Trustees of the Royal Botanic Gardens, Kew, 2005

completely united or rarely free at apex; stigma 5-lobed or globose, rarely 5 separate. Drupelets black, subglobose, 5–8(–10) mm diameter, inserted near the base, strongly wrinkled in dry state. Fig 4 (p. 30).

TANZANIA. Ufipa District: Mbisi, 6 Oct. 1950, *Bullock* 3424! & 13 km on Sumbawanga–Tunduma road, 20 Feb. 1994, *Bidgood et al.* 2354!; Iringa District: Mafinga [Sao Hill], *Shabani* in *Procter* 2710!
DISTR. **T** 4, 7; S Congo (Kinshasa), Angola (Bié), Zambia, and Malawi
HAB. *Uapaca, Brachystegia, Parinari* woodland, open bushland, fireswept eroded hillsides; 1800–2200 m

SYN. *O. humilis* Engl. in E.J. 30: 354, fig. G–K (1901); Gilg in E.J. 33: 233 (1903); T.T.C.L.: 382 (1949). Type: Tanzania, Rungwe District: Upper Kondeland, Umalila, *Goetze* 1353 (B†, holo., BR, EA, iso., K, photo of iso.!), *non* (St. Hil.) Kuntze (1891)
    *O. hockii* De Wild. in Rev. Zool. Afr. 7, Suppl. Bot.: 33 (1919); White, F.F.N.R.: 251, fig. 43j (1962). Type: Congo (Kinshasa), Shaba, Manika, *Hock* s.n. (BR, holo.)

NOTE. Many of the fruiting specimens without rootstocks annotated by me and others as *O. katangensis* could equally well be *O. richardsiae*. I have been reluctant to combine the two. A study in the field of the exact habit and relations to characters of young flowers is needed.

31. **Ochna richardsiae** *Robson* in Bol. Soc. Brot., sér 2, 36: 29 (1962) & in F.Z. 2: 246, t. 45 (1963); Bamps, F.C.B. Ochnaceae: 11 (1967). Type: Zambia, Mbala [Abercorn] District, Ballymain, *Richards* 7017 (K!, holo.)

Shrub 30–60 cm tall with pale brown branches with ± peeling bark; youngest parts with dense raised lenticels the same colour, sometimes papillose-puberulous (not in E Africa?). Leaves ± subcoriaceous, narrowly oblong to narrowly oblanceolate, 1.7–7.5(–12) cm long, 0.6–2 cm wide, acute to narrowly rounded at the apex, cuneate to narrowly rounded at the base, crenate-serrate at the margin, the teeth incurved or directed upwards; midrib, lateral veins and tertiary venation distinctly prominent on both surfaces; petiole short, 1–2 mm long, thick and channelled; stipules triangular, 2–3 mm long, striate, very soon deciduous. Flowers borne on leafless branches, (1–)2–4(–6) on very short spur-shoots ringed with close bract scars; buds globose; pedicels 1–1.5 cm long extending to 1.3–2 cm long in fruit, jointed at base or 3–4 mm from base. Sepals broadly elliptic or round, 5–8 mm long, 4–6 mm wide, becoming red and expanding to (9–)15–17 mm long, 8 mm wide in fruit. Petals yellow, obovate, 9–10(–15) mm long, 6(–14) mm wide. Anthers 2.5 mm long, opening by apical slits; filaments 2 mm long. Carpels 5; styles completely joined; stigma ± capitate. Drupelets green but presumably turning black, ± globose, (5–)8 mm diameter. Fig. 5 (p. 32).

TANZANIA. Iringa District: Mbeya – Iringa road, road to Madibira, 15 Nov. 1966, *Richards* 21589! & 30 km W of Mafinga on the Madibira road by Ndembera R., Penny Penns Farm, 26 Dec. 1986, *Lovett & Congdon* 1176!
DISTR. **T** ?4, 7; Congo (Kinshasa) (Shaba), Zambia, Malawi, Zimbabwe
HAB. Grassland and riverine thicket of *Syzygium guineense* and *Bequaertiodendron magalismontanum*; 1350–1500 m

NOTE. Although there are some small discrepancies with Robson's description e.g. lack of any trace of puberulence on young shoots and petal size, I think the two specimens cited belong here; but see note after *O. katangensis*.

32. **Ochna gambleoides** *Robson* in Bol. Soc. Brot. sér. 2, 36: 30 (1962) & in F.Z. 2: 246 (1963). Type: Zambia, Chadiza turn-off to Fort Jameson, 1.7 km, *Robson* 32 (K!, holo., BM, LISC, PRE, SRGH, iso.)

Small tree 3–7 m tall with slender trunk or sometimes branched from base; bark silver-grey, rectangularly fissured near base, smooth above; branches stout, brown, sometimes lenticellate and often somewhat ridged when young. Leaves glaucous, grey-green in life, tufted at ends of shoots, broadly oblong-elliptic to obovate,

© The Board of Trustees of the Royal Botanic Gardens, Kew, 2005

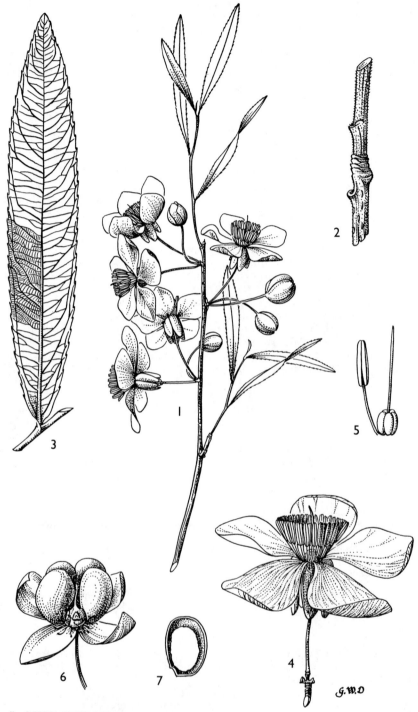

FIG. 5. *OCHNA RICHARDSIAE* — **1**, flowering shoot, × 1; **2**, stem, × 4; **3**, leaf, × 1; **4**, flower, × 2; **5**, stamen and gynoecium, × 4; **6**, fruit, × 4; **7**, section of drupelet, × 4. 1–2, 4–5 from Richards 7017; 3, 6–7 from Fanshawe 1762. Drawn by G.W. Dalby, and reproduced from Flora Zambesiaca.

© The Board of Trustees of the Royal Botanic Gardens, Kew, 2005

4.5–14.5 cm long, 3.5–8.5 cm wide, very broadly rounded or slightly emarginate at apex, broadly cuneate at the base, crenate-serrulate; lateral veins prominent on both surfaces and very closely reticulate, the tertiary venation less so; petiole 5–20 mm long; stipules triangular, 4 mm long, soon deciduous. Fruiting specimens only known, but from this evidence flowers 7–16 in condensed racemiform inflorescences with rhachis up to 13 mm long, several forming a ± globose cluster 8–15 cm wide at end of branches; pedicels 2.6–4 cm long, jointed at base. Fruiting sepals orange-red to scarlet, 16–25 mm long, 10–15 mm wide. Anthers 2 mm long, opening by longitudinal slits (fide Robson). Carpels 5–6, the styles completely united; stigma 5-lobed. Drupelets black, ellipsoid, 10–13 mm long, 8–10 mm wide, inserted at base.

TANZANIA. Mpanda District, 42 km S of Inyonga [Nyonga], 24 Sept. 1961, *Boaler* 333!; Iringa District: Katonda Pass, 64 km on Kilosa road, 20 Oct. 1936, *B.D. Burtt* 6403!
DISTR. **T** 4, 7; Zambia, Malawi, Mozambique and Zimbabwe
HAB. Escarpment *Brachystegia microphylla* woodland and *Brachystegia spiciformis* woodland; ± 1200 m

33. **Ochna schweinfurthiana** *F. Hoffm.*, Beitr. Fl. Centr.-Ost-Afr.: 20 (1889); Gilg in E.J. 33: 234, 240 (1903); T.T.C.L.: 383 (1949); I.T.U. ed. 2: 281 (1952); Keay, F.W.T.A. ed. 2, 1: 223 (1954); Palgrave, Trees Centr. Afr.: 302, illustr. (1956); White, F.F.N.R.: 251, fig. 43h (1962); Robson in F.Z. 1: 248 (1963); Haerdi in Acta Tropica Suppl. 8: 101 (1964); Bamps, F.C.B. Ochnaceae: 15, fig. 2b (1967); Vollesen in Opera Bot. 59: 25 (1980) & Fl. Ethiopia & Eritrea 2 (2): 67, fig. 69.1/1–3 (1995). Type: Tanzania, Tabora District: Igonda [Gonda], *Boehm* 1632 (CORD, lecto., EA, K, photo.!)

Shrub or small tree 2–7(–9) m tall with ± thick dark grey corrugated or reticulately fissured corky bark; slash raw meat coloured; branches mostly not lenticellate, with white or pale yellowish brown bark peeling off in papery strips. Leaves ± coriaceous, obovate-oblanceolate to oblong, 5.5–13.5(–17.5) cm long, (1.3–)1.7–6.5 cm wide, broadly to narrowly rounded at the apex, narrowly cuneate at the base, the margin densely curved-serrulate; lateral veins ± 20, widely spreading, more prominent above than beneath; finely reticulate tertiary venation prominent on both surfaces or less so beneath; petiole (3–)5–12 mm long, thick and flattened above. Flowers precocious, 4–10 in racemiform clusters with rhachis up to 8 mm long or pseudoumbellate; pedicels (1–)1.5–3.5(–4.2) cm long, jointed almost at the base. Sepals obovate-elliptic, 4–6 cm long, 3–4.5 wide, rounded, becoming orange-red to deep red in fruit, 10–15 mm long, 6–8 mm wide. Petals bright yellow, obovate to obovate-oblong, 5.5–10 mm long, 4–5.5 mm wide. Anthers 1–2 mm long, about $^2/_3$ the length of the filaments, dehiscing by longitudinal slits. Carpels usually 5; styles completely united; stigma globose or slightly bilobed. Drupelets black, subglobose, 7–9(–10) mm long, 6–7(–10) mm wide, inserted near the base.

UGANDA. West Nile District: Paidi, Feb. 1934, *Eggeling* 1504! & Uleppi, Mar. 1935, *Eggeling* 1956!; Teso District: Gola Hill, 15 Sept. 1946, *A.S. Thomas* 4546!
TANZANIA. Bukoba District: Nshamba, Sept. 1935, *Gillman* 560a!; Kigoma District: Uvinza, 29 Aug. 1950, *Bullock* 3239!; Chunya District: Lupa N Forest Reserve, 5 Nov. 1963, *Boaler* 1037!
DISTR. **U** 1, 3; **T** 1, 4–8; Mali, Ghana, Togo, Nigeria, Congo (Kinshasa), Rwanda, Burundi, Sudan, Ethiopia, Angola, Zambia, Malawi, Mozambique and Zimbabwe
HAB. *Brachystegia, Julbernardia, Terminalia* etc. woodland and wooded grassland, particularly on rocky hillsides and scarps, *Grewia – Strychnos* etc. scrub and *Oxytenanthera – Hyparrhenia* bushland, also recorded from *Monanthotaxis poggei* riverine forest; 750–2100 m

SYN. *O. suberosa* De Wild. in Rev. Zool. Afr. 7, Suppl. Bot.: 39 (1919). Type: Congo (Kinshasa), Lubumbashi, *Hock* s.n. (BR, holo.)

NOTE. Richards and Arasululu (25832) state young leaves sticky and there do appear to be some minute glands in the areoles beneath.

© The Board of Trustees of the Royal Botanic Gardens, Kew, 2005

34. **Ochna afzelii** *Oliv.* in F.T.A. 1: 319 (1868); Gilg in E.J. 33: 239 (1903); I.T.U. ed. 2: 279 (1951); Keay, F.W.T.A. ed. 2, 1: 223 (1954); White, F.F.N.R.: 251, fig. 43i (1962); Robson in F.Z. 2: 249 (1963); Bamps, F.C.B. Ochnaceae: 13 (1967). Type: Sierra Leone, *Afzelius* s.n. (BM!, holo.)

Shrub or tree (1.5–)3–12 m tall with smooth greybrown sometimes flaking bark; slash meat-red, soft and granular; branches densely lenticellate, purplish or dark brown (rarely whitish). Leaves oblanceolate, oblong-oblanceolate or elliptic, 4–13 cm long, 1.5–4.5 cm wide, ± acuminate to a narrowly rounded tip or rounded at the apex, cuneate at the base, densely curved-serrulate or crenulate at the margin, fairly thin; lateral veins ± 30 with densely reticulate tertiary venation prominent on both surfaces but less so beneath; petiole 2.5–5(–8) mm long, slender, channelled above. Flowers 2–6(–8), pseudo-umbellate or with rhachis up to 4 mm long; pedicels 0.8–3.5 cm long, jointed at or very near the base. Sepals oblong-elliptic, 4–6(–8) mm long, 2–2.5 wide in flower, rounded at the apex, turning scarlet in fruit and enlarging to 6–14(–18) mm long, 3–5.5 mm wide. Petals white to lemon-yellow, obovate, 7–13 mm long, 3–7 mm wide. Anthers orange, 1–1.5(–2) mm long, about $\frac{1}{2}$ as long as the filaments, dehiscing by longitudinal slits. Carpels (5–)6–8 with styles completely united; stigma globose or slightly 5–8-lobed. Drupelets black, subglobose to subreniform, 6–7 mm diameter, inserted below their centre.

subsp. **afzelii**; Robson in F.Z. 2: 249 (1962)

Fruiting sepals narrower and less imbricated, 6–10 mm long, up to 5.5 mm wide; inflorescence rhachis usually shorter; corolla white or pale yellow.

UGANDA. West Nile District: Mt Otzi, 7 June 1936, *A.S. Thomas* 1980!; Mbale District: Mt Elgon, Nabumale, 24 Apr. 1924, *Snowden* 873!; Mengo District: Entebbe, Jan. 1932, *Hancock* in A.D. S2406!
TANZANIA. Bukoba District: Nshamba, Sept. 1935, *Gillman* 560b!; Mwanza District: Rubondo I., N end, 17 Oct. 1985, *FitzGibbon & Barcock* 75 (Q2)!
DISTR. **U** 1, 3, 4; **T** 1, 4 (fide EA); Guinea to Cameroon, Sudan, E Congo (Kinshasa), Angola and Zambia
HAB. Grassland on rocky hills, bushland, evergreen forest; 1150–1350 m

SYN. *O.* sp.; Oliv. in Trans. Linn. Soc. 29: 44 (1872)
    *O. rhodesica* R.E. Fr., Ergebn. Schwed. Rhod.-Kongo-Exped. 1: 149, t. 13, figs 1, 2 (1914). Type: Zambia, Mbala [Abercorn], *Fries* 1248 (UPS, holo.)
    *O. alba* sensu Chev., Bot.: 106 (1920) pro parte; F.W.T.A. ed. 1, 1: 191 (1927) pro parte; I.T.U. ed. 2: 278 (1952)

subsp. **mechowiana** (*O. Hoffm.*) *Robson* in Bot. Soc. Brot., sér. 2, 36: 3 (1962) & in F.Z. 2: 250 (1963). Type: Angola, Malange, *Mechow* 217 (B†, holo.)

Fruiting sepals broader and more imbricated, (10–)12–18 mm long, 9 mm wide; inflorescence rhachis often more developed, corolla lemon yellow to bright yellow.

TANZANIA. Sumbawanga District: Rukwa, 11 km SSW of Tatanda, 24 Oct. 1992, *Gereau & Kayombo* 1205 (fide EA); Mbeya District: Mbosi, Zambi, 16 Nov. 1932, *Davies* 705! & Mbozi [Mbosi], 8 Nov. 1935, *Horbrugh Porter* s.n.!
DISTR. **T** 4, 7; Congo (Kinshasa), Angola, Zambia
HAB. *Brachystegia* woodland; 1550–1650 m

SYN. *O. mechowiana* O. Hoffm. in Linnaea 43: 123 (1881); Gilg in E.J. 33: 234 (1903); R.E. Fr., Schwed. Rhod.-Kongo-Exped. 1: 150, t. 3, fig. 3 (1914); Exell & Mendonça, C.F.A. 1 (2): 288 (1951) pro parte
    *O. welwitschii* Rolfe in Bol. Soc. Brot. 11: 84 (1893); Hiern, Cat. Afr. Pl. Welw. 1: 121 (1896). Type: Angola, Golungo Alto, Monte de Queta, *Welwitsch* 4594 (K!, lecto., BM, COI, LISU, isolecto.)

NOTE. Bamps does not recognise subsp. *congoensis* (Tiegh.) Robson nor subsp. *mechowiana* (O. Hoffm.) Robson but sinks them without comment. There does seem to be a difference between the **T** 7 material and the rest and I have retained subsp. *mechowiana*. Since Robson records both subsp. *afzelii* and subsp. *mechowiana* from Mbala varietal rank might be more appropriate.

© The Board of Trustees of the Royal Botanic Gardens, Kew, 2005

35. **Ochna confusa** *Burtt Davy & Greenway*, Fl. Pl. & Ferns Transvaal 1: 238, 239 (1926) & in K.B. 1926: 239 (1926); Robson in F.Z. 2: 250 (1963); du Toit & Obermayer, F.S.A. 22: 7 (1976). Type: South Africa, Gauteng Lydenberg, Pilgrims Rest, *Rogers* 23068 (K!, holo., PRE, iso.)

Shrub 0.1–1(–2) m tall or shrublet with woody rootstock; bark brown becoming vertically striate; branches virgate, yellowish white, densely pustulose-lenticellate. Leaves ± thin, elliptic to oblong-elliptic or oblanceolate, 4.4–9(–11) cm long, (1–)1.5–2.1(–2.9) cm wide, acute to subacuminate at the apex, cuneate at the base, margin densely serrulate with straight or incurved teeth; lateral veins and very densely reticulate tertiary venation prominent above, less so beneath; petiole 2–4 mm long; stipules oblong-lanceolate, naviculate, 5–7 mm long, sometimes somewhat persistent. Flowers ± precocious, (2–)3–6 in very short racemiform or pseudoumbellate axillary inflorescences; pedicels 0.9–1.2 cm long, jointed at the base, sometimes obscurely muriculate. Sepals oblong-elliptic, 4–5(–7) mm long, 3–4 mm wide, rounded, becoming red in fruit and 8–11 mm long, 5–7 mm wide. Petals bright yellow, broadly elliptic or ± round, 7–10 mm long, 3–4(–8 annot. by Robson) mm wide. Anthers 1–1.5 mm long, $^1/_3$–$^1/_2$ as long as the filaments, dehiscing by longitudinal slits. Carpels 5, the styles completely united; stigma 5-lobed. Drupelets subglobose, 6–8 mm long, 5–6.5 mm wide.

TANZANIA. Mbeya District: 10 km NW of Tunduma, 15 Nov. 1986, *Brummitt et al.* 17982!; Iringa District: 48 km S of Iringa on Mbeya road, 14 Nov. 1958, *Napper* 864!; Njombe District: Njombe, 29 Nov. 1931, *Lynes* D57 (fl.)! & D27 (fr.)!
DISTR. **T** 7; Zambia, Malawi, Mozambique, Zimbabwe, N South Africa
HAB. *Brachystegia* woodland; 1450–1750 m

SYN. *O. gracilipes* sensu Brenan in Mem. N.Y. Bot. Gard. 8: 234 (1953), *non* Hiern

NOTE. Robson (F.Z. 2: 251 (1963)) points out that this may not be specifically distinct from *O. pygmaea* Hiern.

36. **Ochna** sp. 36

Shrub or small tree with silvery grey trunk; branches dull greyish with pale unraised lenticels, with numerous short side branches. Leaves completely lacking at flowering time. Flowers 4–9 in very short inflorescences borne on nodular spur shoots, strongly violet-scented; pedicels slender, 4–9.5 mm long, jointed 1.5–3 mm from base. Sepals elliptic, 7 mm long, 3.5 mm wide. Petals pale canary yellow, obovate, 6.5 mm long, 2.5 mm wide. Anthers 1.2 mm long, opening by longitudinal slits, about equalling the filaments. Stigma capitate. Fruits not known.

TANZANIA. Dodoma District: Manyoni, Kazikazi, 23 Dec. 1932, *B.D. Burtt* 3831 p.p.!
DISTR. **T** 5; not known elsewhere
HAB. Itigi type deciduous thicket; ± 1260 m

NOTE. Robson showed that part of one of the two sheets of 3831 at Kew was *Ochna ovata* and the other part he annotated "Ochna sp. longitudinal dehiscence, paniculate inflorescence". Burtt's field note "common in the thicket, fls. appear with first rains and only last a day" probably refers to this since it forms the bulk of the material on the two sheets. It is just possible that *Burtt* 3932, a sterile leafy specimen from Mpwapwa, 17 Apr. 1932, dry thicketed hill slope at 1200 m can be associated with it. It has oblong-oblanceolate acute or shortly acuminate leaves up to 10 × 3.5 cm.

37. **Ochna** sp. 37

Much-branched small tree; branches dark purplish with pale lenticels; apical buds of triangular striate scales 3 mm long. Leaves oblanceolate to oblanceolate-elliptic, 3.5–13.5 cm long, 1–4.2 cm wide, sharply acuminate at the apex, narrowly cuneate at the base and decurrent on to petiole as raised ridges, margin serrulate-spinulose;

© The Board of Trustees of the Royal Botanic Gardens, Kew, 2005

lateral veins ± 35, curving upwards at margins, ± prominent on both sides; tertiary venation densely minutely reticulate beneath, ± plane but more open above and prominent; petiole 3–5 mm long; stipules narrowly triangular, 2 mm long. Flowers and fruits unknown.

TANZANIA. Rufiji District: Mafia I., Kikuni, 13 Aug. 1937, *Greenway* 5085!
DISTR. **T** 6; not known elsewhere
HAB. Evergreen forest of *Diospyros, Mimusops, Ricinodendron* and *Markhamia zanzibarica* on coral rock; ± 9 m

NOTE. Specimens of this were sent to Berlin, where it was considered a new species; no other material has been seen. It is only mentioned to draw attention to the need for adequate material but it is probably already extinct.

### 38. **Ochna** sp. 38

Shrub ± 3 m tall with very slender branches and numerous short leafy side shoots; lenticels obscure. Leaves thin, elliptic, 1.2–4.5 cm long, 0.5–2 cm wide, sharply acute at the apex, cuneate at the base, finely setulose-serrate at the margins; midrib pale yellowish and prominent on both surfaces; lateral veins ± 15, rather irregular together with closely reticulate tertiary venation, finely prominent on both surfaces; petiole 1.5 mm long. Flowers solitary on short leafy shoots but not subtended by a leaf, not seen in young state; pedicels reddish, 10–13 mm long, jointed about 1 mm from base, thickened beneath calyx. Fruiting sepals bright red, narrowly-elliptic, 9–11 mm long, 5 mm wide, spreading. Drupelets not seen.

TANZANIA. Lindi District: Rondo Plateau, western edge of Rondo Forest Reserve, 19 Feb. 1991, *Bidgood et al.* 1661!
DISTR. **T** 8; not known elsewhere
HAB. Semi-evergreen forest of *Milicia, Albizia, Dialium, Pteleopsis* etc. on grey sandy soil; ± 750 m

### 39. **Ochna** sp. 39

Small tree to 3 m with smooth grey bark; young branches blackish purple, the youngest shoots without visible lenticels but older ones with numerous pale lenticels. Leaves thin, oblanceolate to obovate or ± elliptic, 2.5–6 cm long, 1.2–3 cm wide, subacute and minutely apiculate at the apex, cuneate at the base, serrulate with small aculeolate incurved teeth; lateral veins ± 25, together with closely reticulate venation, distinctly prominent on both surfaces; petiole 2 mm long; stipules 2–3 mm long, striate with short filiform fimbriate at apex. Flowers solitary; pedicels 1–2.3 cm long, jointed 2–4 mm from the base. Young flowers not known; fruiting sepals red beneath, green above, oblong-elliptic, 12 mm long, 6–8 mm wide. Styles united but distinctly free at apex, the branches 1.5–2 mm long; stigmas minute. Drupelets black, ellipsoid, 8 mm long, 6 mm wide, attached distinctly above the base.

TANZANIA. ?Kilosa District: R. Ruaha, 10 Jan. 1956, *Benedicto* 100! & Ruaha R. gorge E of Ruaha, 8 Jan. 1975, *Brummitt & Polhill* 13612!
DISTR. **T** 6; not known elsewhere
HAB. Deciduous bushland of *Adansonia, Cordyla, Euphorbia, Acacia, Afzelia* etc.; 500–600 m

NOTE. The first cited sheet had been determined as *O. ovata* F. Hoffm. but differs from that in that the leaves are more narrowed at the base, flowers uniformly solitary, fruiting sepals larger.

© The Board of Trustees of the Royal Botanic Gardens, Kew, 2005

### 40. **Ochna** sp. 40

Tree to 15 m with smooth bark and pinkish brown slash. Leaves rather thin, narrowly elliptic to oblanceolate, 6–8 cm long, 1.6–2.5 cm wide, acuminate at the apex, cuneate at the base, margin closely serrulate with curved aculeolate teeth; lateral veins ± 15, irregular, ascending, together with the reticulate venation, prominent on both surfaces. Flowers 2–5 at the end of short spur shoots; axis of inflorescence ± 5 mm long; young flowers not seen, pedicels of young fruits up to 2.5 cm long; jointed 1–3 mm from base. Enlarged sepals pinkish red, imbricate, ± round, or broadly ovate, 12 mm long and 11–12 mm wide.

UGANDA. Ankole District: Bushenyi, S Kasyoha – Kitomi Forest, Kamukaaga, between R. Kyambura and R. Nzozi, Oct. 1998, *Hafashimana* 0662!
DISTR. **U** 2; not known elsewhere
HAB. Mature moist evergreen forest, *Parinari – Strombosia* etc.; ± 1400 m

NOTE. This had been named *O. holstii* and confirmed as such by Wadhwa; the habit and habitat agree with that species but the enlarged sepals are more imbricate and much broader and I have hesitated to include it in that species. The forest is known for endemism.

### 41. **Ochna** sp. 41

Shrub or small tree to 4.5 m tall with very slender black branches with pale lenticels. Leaves thin, glossy, narrowly ovate-lanceolate to lanceolate, 2–6.5 cm long, 0.7–1.8 cm wide, finely acute at the apex, ± rounded at the base, slightly or very inconspicuously toothed at margin, often appearing almost entire; lateral veins numerous, prominent above but tertiary venation less discernable than usual; petioles slender, 2.5 mm long. Inflorescence buds very slender, 5 mm long, 1.5 mm wide. Flowers 2–3 on very short spur shoots, not known in young state; pedicels 2.5–3 cm long, jointed ± 4 mm from the base. Fruiting sepals 8 mm long, 5 mm wide. Drupelets ellipsoid, 9.5 mm long, 7 mm wide.

TANZANIA. Kilosa District: Mikumi National Park, Mvuma Hill area, 24 km from HQ, 6 July 1973, *Greenway & Kanuri* 15367! & Mikumi National Park, Mkata R., Dec. 1967, *Procter* 3797!
DISTR. **T** 6; not known elsewhere
HAB. Riverine forest, *Erythroxylum*, *Byrsocarpus*, *Oncoba*, *Fagara* thicket, *Afzelia*, *Terminalia* and *Sterculia appendiculata*; ? ± 500 m.

NOTE. *Procter* 3797 had been named *O. oxyphylla* but is from too low an altitude, has glabrous stems and joint of pedicel is too long. Nothing can be done without adequate flowering material; I have been unable to match it with any known species.

### 42. **Ochna** sp. 42

Tree 4 mm tall with thick brown bark; very young leafy branchlets short, reddish brown, older branchlets thick with corky fissured bark, orange-brown powdery beneath. Leaves fairly thin, narrowly elliptic-oblong to ± oblanceolate, 6.5–11 cm long, 2–3.2 cm wide, rounded or very slightly emarginate at the apex, narrowed to rounded base, serrate-crenulate at margin; lateral veins 20–25, rather widely spaced, together with the extensive reticulate tertiary venation, prominent on both surfaces; petiole thick, 2 mm long. Flowers paired on very short spurs; pedicels 1.6 cm long, jointed 4 mm from base. Fruiting sepals ± 8 mm long and wide. Young flowers not known.

TANZANIA. Uzaramo District: Pande Hill, 22 Nov. 1969, *Harris et al.* 3619!
DISTR. **T** 6; not known elsewhere
HAB. Wooded grassland; ± 75 m

© The Board of Trustees of the Royal Botanic Gardens, Kew, 2005

### 43. Ochna sp. 43

Shrub or small tree to 3 m tall; branches ridged, young parts pale brown with dense raised lenticels the same colour. Leaves ± coriaceous, elliptic to elliptic-oblanceolate, 2–7.5 cm long, 1–2.8 cm wide, acute to shortly acuminate at the apex, cuneate at the base, rather coarsely crenate-serrate, the teeth strongly curved inward so that margin appears simply crenate; midrib strongly prominent but other venation only slightly so; petioles thick 2–3 mm long; stipules linear-lanceolate, up to 12 mm long, soon deciduous. Flowers not known in young state, solitary or paired on short spur-shoots; pedicels 12–20 mm long, jointed 1–5 mm from base. Fruiting sepals red, ovate-elliptic, ± 10 mm long, 8 mm wide. Drupelets black, ellipsoid, 9 mm long, 6 mm wide.

TANZANIA. Iringa District: dam end of Lake Ngwazi, 20 Oct. 1987, *Lovett & D.W. Thomas* 2408!
DISTR. **T** 7; not known elsewhere
HAB. Stunted *Brachystegia* woodland; ± 1800 m

### 44. Ochna sp. 44

Probably tree (no data); branchlets blackish purple, densely pale lenticellate, the younger parts minutely papillate. Foliage drying brown, paler beneath; leaves elliptic, narrowly elliptic or elliptic-oblong, 3.5–11.5 cm long, 1.2–4 cm wide, subacute at apex, the extreme tip shortly apiculate, cuneate to narrowly rounded at base, serrulate with incurved teeth at margin; lateral veins ± 30–40, together with subparallel and reticulate tertiary veinlets prominent on both surfaces; petiole 3–3.5 mm long, winged by decurrent leaf-base; stipules short, triangular, ± 1 mm long, sometimes fimbriate at apex. Flowers 2–4 on very short spur shoots annulated by fallen bracts; pedicels 1.3–2.5 mm long, jointed 4–8 mm from base, sometimes minutely papillate. Fruiting sepals elliptic, 8 mm long, 3–4 mm wide; young flowers not known. Drupelets ellipsoid–slightly reniform, 7–8 mm long, 4–5 mm wide.

TANZANIA. Mwanza District: presumably on or near Ukerewe I., Apr. 1928, *Conrads* 6041! and EA 13307! & 24 Nov. 1932, *Conrads* EA 13312!
DISTR. **T** 1; not known elsewhere
HAB. Not known

NOTE. Without better material with young flowers I have been unable to name this. A completely sterile specimen *Greenway & Kanuri* EA 13981, Mar. 1968, a 3 m shrub from granite inselbergs at **T** 1, Maswa District: Handajega (said to equal *Conrads* 5520 which I have not seen) may be the same with more rounded leaf-bases.

### 45. Ochna sp. 45

Shrub 0.5 m tall; young stem greyish white, rugose. Leaves fairly thin, narrowly oblong-oblanceolate, 6–11 cm long, 1.5–2.5 cm wide, narrowly acute to acuminate at the apex, narrowly cuneate at the base, margin closely serrulate with short ± erect dark-tipped teeth; lateral veins ± 20, curving upwards with numerous rather irregular intermediaries and closely reticulate tertiary venation all ± prominent on both surfaces; petiole 4–5 mm long. Inflorescence borne at end of short shoot; axis about 1 cm long with 5–6 close 2–4-flowered cymes with about ± 25 flowers judging by pedicel remnants; pedicels ± 15 mm long with joint up to 5 mm from base. Fruiting sepals poorly preserved, probably 8 mm long, 2–5 mm wide. Drupelets black, subglobose, 8 mm long, 6.5 mm wide.

TANZANIA. Rufiji District: Utete, mouth of Rufiji Delta, Mchungu Forest, 23 Aug. 1990, *Frontier Tanzania* 1370!
DISTR. **T** 6; not known elsewhere
HAB. Coastal forest on sand with dense evergreen undershrubs; near sea-level

© The Board of Trustees of the Royal Botanic Gardens, Kew, 2005

46. **Ochna** sp. 46

Tree to 6 m high. Leaves lanceolate, 5–9.5 cm long, 2–2.5 cm wide, narrowly long-acuminate at apex, cuneate at base, closely toothed at the margin; petioles 2 mm long; stipules pale brown, linear, 11–13.5 mm long, ± chaffy, very deciduous. Inflorescences at least 9-flowered, on short shoots; rhachis 5 mm long; pedicels 15 mm long, jointed close to base (save for apical flower). Sepals 7 mm long, 1.8–3.5 mm wide. Filaments ± 2.5 mm long; style 4.5 mm long; stigma capitate.

TANZANIA. Iringa District: Udzungwa Mountain National Park, 07°46' S 36°50' E, 28 Sep. 2001, *Luke et al.* 7919!
DISTR. **T** 7; not known elsewhere
HAB. Montane forest; ± 1700 m

NOTE. This had been first tentatively determined as *O. afzelii* and then as *O. holstii* but the stipules fit neither species, but resemble those of species 43 – though that has different leaves and inflorescences.


47. **Ochna** sp. 47

Tree to 20 m high with pale redbrown bark which peels easily. Leaves drying dark, elliptic, 7 cm long, 2–3 cm wide, narrowly acuminate at the apex, cuneate at the base, crenulate, each crenulation with incurved tooth; midrib drying blackish beneath; lateral veins not very clearly differentiated from tertiary reticulate venation; petiole 2 mm long, drying black. Flowers solitary at apices of slender side branches 2–3 cm long; pedicels 2–3 cm long, jointed about 1 mm from base. Sepals oblong-elliptic, 10–12 mm long, 5–6 mm wide; petals bright yellow, rounded or obovate, 10–12 mm long, 6–7 mm wide with narrow claw. Anthers 2.2–2.5 mm long, opening by longitudinal slits; filaments 2–3.5 mm long. Carpels 6; style 6–8 mm long; stgma capitate, ± lobed. Drupelets not seen.

TANZANIA. Morogoro District: Uluguru South catchment Forest Reserve, W slopes on the path from N'gungulu village to Lukwangule Plateau, 5 Feb. 2001, *Jannerup & Mhoro* 405!
DISTR. **T** 6; not known elsewhere
HAB. Open ridge forest with *Cussonia*, *Garcinia* and *Lasianthus*; ± 2400 m

NOTE. This had been determined as *O. holstii* but that has several-flowered inflorescences with a distinct rachis. No *O. holstii* has been seen from the Uluguru Mts proper.


ERRONEOUS RECORD

On 18 Apr. 1909 Baroa de Soutellinho sent to Kew a small specimen of an *Ochna* he had cultivated in Oporto from seeds said to be from "Ruwenzori, Central Africa". Robson has identified this as the South African species *O. atropurpurea* DC. and the seeds could not have come from East Africa.


## 2. BRACKENRIDGEA

A. Gray, Bot. U.S. Expl. Exped. 1: 361, t. 42 (1854)

Glabrous trees, shrubs or subshrubs with yellow pigment under the bark. Leaves petiolate; lamina entire to glandular-serrate with characteristic venation; stipules free, longitudinally striate, laciniate or deeply divided into linear segments, often persistent on young shoots. Flowers solitary or in panicles or fascicles, the fascicles sometimes forming spikes or heads, terminal or at base of young growth. Sepals (4–)5, usually quincuncially imbricate in bud, persistent and becoming red and coriaceous in fruit. Petals (4–)5, white to pink, deciduous. Stamens (8–)10–20(–22),

© The Board of Trustees of the Royal Botanic Gardens, Kew, 2005

free; anthers yellow, dehiscing by longitudinal slits, deciduous; filaments slender, persistent. Carpels (3–)5–10, free at base, 1-ovuled; styles slender, gynobasic with small stigmas. Fruit of 1-several free drupelets with fleshy mesocarp and with internal projections of endocarp. Seeds curved, without endosperm.

A genus of about a dozen species in Africa, Madagascar and Malaysia from Andaman Is, Perak and the Philippines to New Guinea, also Fiji and Australia (Queensland).

1. Mostly a tree 5–10 m tall, only infrequently a shrublet 0.3–2 m; stamens 10–11; petals 4–5 mm long; sepals white or greenish in flowering state; more coastal in distribution (**K** 7, **T** 3, 6, 8) . . . . . . . . . . . . . . . . . . . . . .     1. *B. zanguebarica**
2. Always a rhizomatous shrublet or shrub 0.05–1(–2) m tall, never a tree; stamens 13–22; petals 6–8 mm long; sepals pink in flowering state; more western in distribution (**T** 1, 4) . . . . . . . . . . . . . . . . . . . . . . . . . . .     2. *B. arenaria*

1. **Brackenridgea zanguebarica** *Oliv.* in Hook. Ic. 11: 77, t. 1096 (1871); V.E. 3 (2): 486 (1921); T.T.C.L.: 381 (1949); K.T.S.: 336 (1961); Robson in F.Z. 2: 252, t. 46a (1963); Haerdi in Acta Trop., Suppl. 8: 100 (1964); Vollesen in Opera Bot. 59: 25 (1980); K.T.S.L.: 121, fig. (1994). Type: Tanzania, Uzaramo District: Dar es Salaam, *Kirk* 140 (K!, holo. & iso.)

Shrub or small tree (0.3–)2–10(–15) m tall, deciduous, with tough or deeply furrowed black or dark grey corky bark, yellow beneath; branches at first purplish with pale lenticels, later whitish. Leaves oblong or elliptic to oblanceolate or obovate, 2.5–7.7(–11.8) cm long, (0.9–)1–2.6(–3.6) cm wide, acute or shortly mucronate to rounded at the apex, cuneate at the base, densely glandular-serrate, the teeth often a characteristic orange-brown; venation with primary lateral veins ascending, with more widely spreading secondary laterals and tertiary cross veins, prominent above; petiole 1–2.5 mm long. Flowers appearing before the leaves or when leaves are young or few, solitary or 2.4(–8) in fascicles, axillary or terminating short shoots, scented; pedicels 1–2 cm long. Sepals white or greenish, oblong, 3–4(–5) mm long, rounded, reflexed after flowering and becoming crimson, enlarging to 7–9(–10) mm in fruit and more spreading. Petals white or cream, oblong-lanceolate, 4–5 mm long, narrowed to base. Stamens 10–11; anthers yellow, 1.5–2.5 mm long. Carpels 5; styles completely united. Drupelets black, 6–7 mm long and wide. Fig. 6 (p. 41).

KENYA. Kwale District: Kwale–Tanga road, 38 km from Mombasa, 25 Jan. 1961, *Greenway* 9802! & Shimba Hills, Mwele Mdogo forest, 19 km SW of Kwale, 4 Feb. 1953, *Drummond & Hemsley* 1107!; ?Mombasa District: mainland behind Mombasa I., Apr. 1876, *Hildebrandt* 1966!
TANZANIA. Tanga District: Nyamaku, 27 Jan. 1957, *Faulkner* 1933!; Morogoro District: Morogoro, 16 Dec. 1934, *E.M. Bruce* 357!; Tunduru District: about 1.5 km E of R. Mawese; 18 Dec. 1955, *Milne-Redhead & Taylor* 7691!
DISTR. **K** 7; **T** 2 (fide EA), 3, 6, 8; Mozambique, Malawi, Zimbabwe
HAB. Coastal bushland and wooded grassland, *Brachystegia* woodland, lowland *Julbernardia* evergreen forest; 0–1050 m (see note)

SYN. "*Ochna* (in fruit) and *O. leptoclada* Oliv. ?" Oliv. in Trans. Linn. Soc. Ser. 2, Bot.: 330 (1887)
     *Ochna alboserrata* Engl. in E.J. 17: 75 (1893). Type: Kenya, mainland near Mombasa, *Hildebrandt* 1966 (B†, holo., K!, iso.)

* A single gathering from **T** 6, Mafia I, is evergreen whereas all other material is deciduous and has been treated as 3. *B.* sp.

© The Board of Trustees of the Royal Botanic Gardens, Kew, 2005

FIG. 6. *BRACKENRIDGEA ZANGUEBARICA* — **1**, fruiting branch, × 1; **2**, leaf part, × 8; **3**, flower, × 4; **4**, section of drupelet, × 4. *BRACKENRIDGEA ARENARIA* — **5**, stipule, × 4; **6**, flower, × 4. 1–3 from *Faulkner* 93, 4 from *Gomes e Sousa* 2204; 5 from *Gilges* 245, 6 from *Holmes* 1181. Drawn by G.W. Dalby, and reproduced from Flora Zambesiaca.

© The Board of Trustees of the Royal Botanic Gardens, Kew, 2005

*Brackenridgea bussei* Gilg in E.J. 33: 273 (1903); V.E. 3 (2): 486 (1921); T.T.C.L.: 381 (1949). Types: Tanzania, Kilwa District: Donde-Land near Kwa Mpanda, *Busse* 635 (B†, syn., EA!, isosyn.) & Kilwa District: Njenje [Djenye] stream, *Busse* 656 (B†, syn., EA!, isosyn.)

*Ochna praecox* Sleumer in F.R. 39: 277 (1936); T.T.C.L.: 383 (1949). Type: Tanzania, Lindi District: Lake Lutamba, *Schlieben* 5712 (B†, holo. )

NOTE. Robson gives up to 1525 m in E Africa but this is based on inadequate labels. *Johnston* s.n., Kilimanjaro 5000', was not collected on the mountain but on the journey from the coast. *Wigg* 381 a duplicate at K from Imperial Forestry Institute, very inadequately labelled, states Lushuto but certainly was not collected there.

2. **Brackenridgea arenaria** (*De Wild. & Dur.*) *Robson* in Bol. Soc. Bot., Sér. 2, 36: 37 (1962) & in F.Z. 2: 254, t. 46b (1963); du Toit & Obermayer, F.S.A. 22: 5 (1976). Type: Congo (Kinshasa), Kisantu, *Gillet* 68 (BR, holo.)

Subshrub 0.05–1(–2) m tall from an often branched rhizome; bark brown or greyish, ± rough, flaking; branches purplish with pale lenticels, becoming paler. Leaves petiolate; lamina oblong or elliptic to oblanceolate, less often obovate, 2.5–12(–16) cm long, 1.2–3.7 cm wide, acute or mucronate to rounded at the apex, cuneate at the base, margin glandular-serrate or rarely entire with venation similar to *B. zanguebarica*; petiole 1–4(–14) mm long; stipules laciniate or fin-like, persistent on young stems. Flowers solitary or 2–4 in fascicles, the fascicles often clustered together at the base of young shoots, appearing before the leaves; pedicels (0.9–)1.3–2.5 cm long. Sepals pink, oblong, at first 5–6 mm long, rounded, spreading or reflexed after flowering, becoming crimson-red in fruit and enlarging to 8–10 mm. Petals white, often tinged pink or with a pink central line, obovate to narrowly-oblanceolate, 6–8 mm long, narrowed to the base. Stamens (13–19)20(–21–22); anthers yellow, 1.5–2 mm long, often twisting spirally after dehiscence. Carpels 5–7 with styles completely united; stigma small. Drupelets black, 6–8 mm long and wide.

TANZANIA. Biharomulo District: Bwanga, 10 Oct. 1960, *Tanner* 5265!; Mwanza District: Vuziligembe Forest, Nov. 1967, *Procter* 3776!; Tabora District: Urambo, 30 Dec. 1950, *Moors* 43!
DISTR. T 1, 4; Congo (Kinshasa), Burundi, Angola, Zambia, Zimbabwe, N Namibia
HAB. *Brachystegia* woodland, grassland; 1150–1200 m

SYN. '*Ochna* sp.'; Oliv. in Trans. Linn. Soc. Lond. 29: 43 (1873) (*Grant* s.n. from Tabora District: Tabora [Kazeh] )
O. *ferruginea* Engl. in E.J. 17: 76 (1893), *non* Kuntze (1891). Types: Tanzania, Biharamulo/Mwanza District: Uzinza [Usinja], near Bumpéke, *Stuhlmann* 837 (B†, syn.) & District unclear, Njakamaga, *Stuhlmann* 863 (B†, syn.)
O. *floribunda* Bak. in K.B. 1895: 289 (1895), *non* Kuntze (1891). Type: Zambia, near Mweru, *Carson* 8 (K!, holo.)
O. *arenaria* De Wild. & Dur. in Bull. Herb. Boiss. ser. 2, 1: 7 (1900); Gilg in E.J. 33: 232 (1903); Z.A.E. 2: 10 (1922); Bamps, F.C.B. Ochnaceae: 5, t. 1 (1967). Type as for *B. arenaria*
O. *angustifolia* Engl. & Gilg in E.J. 32: 135 (1902) & in Warb., Kunene–Samb.-Exped.: 304 (1903); Gilg in E.J. 33: 232 (1903); Exell & Mendonça, C.F.A. 1 (2): 285, t. 12g (1951); White, F.F.N.R.: 250, fig. 43d (1962). Type: Angola, Huila, *Antunes* 137 (B†, holo., BM!, fragment, COI, iso.)
*Brackenridgea ferruginea* (Engl.) Tiegh. in Journ. de Bot. 16: 47 (1902); Gilg in E.J. 33: 273 (1903); T.T.C.L.: 381 (1949)
*Ochna roseiflora* Engl. & Gilg in Warb., Kunene–Samb.-Exped.: 304 (1903); Gilg in E.J. 33: 273 (1903). Type: Angola, Bié, between Quipungo and R. Cuetei, *Baum* 813 (B†, holo.)
O. *bequaertii* De Wild. in Rev. Zool. Bot. Afr. 7 Suppl. Bot.: 30 (1919). Types: Congo (Kinshasa), Sankisia, *Bequaert* 179; Manika, *Hock* s.n.; Baala–Kansomma, *Lescrauwaet* 157 (all BR, syn.)

© The Board of Trustees of the Royal Botanic Gardens, Kew, 2005

### 3. **Brackenridgea** sp.

Evergreen shrub to 3 m tall; stems with ridged dark grey bark, lenticellate. Leaves elliptic, 5–9.5 cm long, 1.5–3.5 cm wide, ± acuminate at the apex, cuneate at the base, subcoriaceous, closely glandular-serrate; petiole 4–8 mm long, rather thick and channelled. Flowers green in small axillary clusters with close bract scars at base; pedicels ± 5 mm long. Calyx lobes oblong-elliptic, 5–5.5 mm long, 2–2.5 mm wide, rounded. Petals not seen. Stamens 10; anthers 2.2 mm long, filaments 1.3 mm long. Fruits not seen.

TANZANIA. Rufiji District: Mafia I., Chunguruma, 1 Oct. 1937, *Greenway* 5355!
DISTR. **T** 6; not known elsewhere
HAB. Rare in *Mimusops, Carissa, Salacia, Landolphia*, Icacinaceae, *Eugenia* evergreen bushland on white sandy soil; ± 9 m

NOTE. I am not satisfied that this is merely an evergreen form of *B. zanguebarica*. The gathering of Greenway is all I have seen from the island and the habitat is now unlikely to be extant.

### 3. **GOMPHIA**

Schreb., Gen. Pl. 1: 291 (1789); Kanis in Taxon 16: 420 (1967) & in Blumea 16: 51 (1968) & in Rev. Fl. Ceylon 6: 251 (1988)

*Meesia* Gaertn., Fruct. 1: 344 (1788) *nom. rejic., non* Hedw. (1801) *nom. cons.*
*Campylospermum* Tiegh. in J. Botanique (Morot) 16: 40, 194, 197 (1902); Perrier in Fl. Madag. 133 Ochnacées: 2–18 (1951); Farron in F.C.B. Ochnaceae: 32 (1967), *nom. illeg.*
*Rhabdophyllum* Tiegh. in Bull. Mus. Hist. Nat. Paris 8: 216 (1902); Farron in F.C.B. Ochnaceae: 23 (1967)
*Idertia* Farron in Bull. Soc. Bot. Suisse 73: 212 (1963) & in F.C.B. Ochnaceae: 22 (1967)
*Ouratea* sensu auctt. Afr., *non* Aubl. (1775)

Evergreen trees or shrubs, usually completely glabrous with thin to ± coriaceous, sessile to petiolate leaves with often fine close secondary and tertiary venation; stipules entire, not striate, free or often ± united, persistent or deciduous. Inflorescences paniculate or raceme-like or less often reduced to 1–2 flowers, terminal or axillary at the base of the new growth; bracts deciduous or ± persistent; pedicels usually slender, articulated at or above the base. Sepals 5, quincuncially imbricate, persistent and turning red and coriaceous in fruit. Petals 5, yellow, cream or white, not clawed. Stamens (8–)10 with filaments much shorter than the anthers or ± absent. Carpels 5, ± free at base, 1-ovulate; styles slender, gynobasic, completely united; stigmas terminal, punctiform. Drupelets usually black at maturity, 1–5. Seeds without endosperm; embryo ± curved, incumbent or accumbent, isocotylous or heterocotylous.

A genus of about 60 species mostly in Africa and Madagascar with one in Asia and Malesia. I agree with Kanis and Farron that *Ouratea* must be restricted to the New World but do not believe that the African species should be distributed amongst three genera as accepted by Farron. Van Tieghem kept up a much larger number of genera but I have followed Kanis in recognising only *Gomphia* for the Old World species. Vollesen in Fl. Ethiopia & Eritrea 2 (2): 69 (1995)) takes the same view. *Rhabdophyllum* Tiegh. with its characteristic very close lateral veins is admittedly distinctive but no species of this group have been found in East Africa. In the Index Kewensis *Gomphia* was used for both Old and New World species in preference to the older name *Ouratea* and Stapf and others were still using it in the first decade of the 20th Century despite the fact that Engler had correctly used the older name in Flora Brasiliensis in 1876.

The species are not too difficult to name by comparison with named material but it is not easy to convey the differences in a key. Geography is very helpful.

© The Board of Trustees of the Royal Botanic Gardens, Kew, 2005

1. Inflorescences sessile, only 1–3-flowered, axillary (**U** 2,
   Budongo) ........................................  9. *G. mildbraedii*
   Inflorescences usually pedunculate and with well developed
   axes and more flowers  ........................................ 2
2. Leaves distinctly broadly oblanceolate with some smaller
   ones elliptic or oblong-elliptic, often large, up to 33 cm
   long, 12 cm wide, mostly shortly acuminate to acute at
   the apex; lateral veins 13–20 lying in depressed hollows
   above but actual veins prominent, very prominent
   beneath; supernumerary striate persistent stipules often
   present on internodes; inflorescences up to 35 cm long,
   mostly unbranched (**K** 7, **T** 3, 6, 7) .................  6. *G. sacleuxii*
   Leaves not distinctly oblanceolate, more symmetrical and
   broadest about the middle or if somewhat oblanceolate
   then with more numerous rather obscure lateral veins
   above and rounded at apex ........................................ 3
3. Main lateral veins 10–16, spaced, strongly curved upwards
   and with numerous lesser ones between more or less at
   right angles to midrib; leaves not over 6.5 cm wide, acute
   or acuminate at apex ........................................ 4
   Main lateral veins (13–)20–35, not or not so strongly
   curved upwards or sometimes a few doing so; leaves up
   to 37 cm long and 12.5 cm wide usually rounded at apex
   (Eastern Arc forest species 5, *G. scheffleri* from **K** 7, **T** 3,
   6, 7, with leaves 4–20 × 2–7 cm, bluntly acute, obtuse or
   rounded and 13–20 main veins; sparsely branched or ±
   unbranched, comes here) ........................................ 7
4. Inflorescences condensed, ± subspherical, about 6 × 5 cm
   without clear long divaricating axes; leaves narrowly
   acute to a fine point (restricted to **T** 4, Kigoma and
   Mpanda districts) ........................................  3. *G. lunzuensis*
   Inflorescences not so condensed and with more distinct
   often long divaricate branched axes ........................................ 5
5. Leaves subcoriaceous, more oblong, not so sharply
   pointed; mostly swamp forest species (**U** 4 (Masaka), **T** 1
   (Bukoba) ........................................  4. *G. vogelii*
   Leaves mostly thinner, more narrowly elongate-elliptic or
   lanceolate-elliptic, acute to distinctly acuminate; not so
   restricted to very wet forest ........................................ 6
6.*Inflorescences more open, up to 20(–30) cm long with
   often much fewer (sometimes only 2–3) individual
   branches up to 16 cm long; leaf-margins subentire to
   finely and regularly serrate (**U** 2; **K** 7; **T** 3, 6, 7) ......  1. *G. reticulata*
   Inflorescences more condensed, usually under 10 cm long
   (but can reach 18) with much shorter individual
   branches usually under 10 cm long; leaf-margins closely
   serrate with very evident curled aculeate teeth (**U** 2, 4;
   **K** 5; **T** 1 (Mwanza) ........................................  2. *G. likimiensis*
7. Inflorescences much branched, more dense and with
   many flowers on the branches; leaves quite coriaceous
   and can attain 37 cm long, 12.5 cm wide, the venation
   rather obscure (**U** 2, 4; **K** 5; **T** 1, 4) .................  7. *G. densiflora*
   Inflorescences usually simple axillary pseudoracemes or
   much less branched with fewer flowers born more
   towards apex of branches ........................................ 8

* If in doubt use geography.

© The Board of Trustees of the Royal Botanic Gardens, Kew, 2005

8. Axes of inflorescence rather stout, erect with nodes
　　bearing distinct imbricate crowded bracts; leaves
　　coriaceous, up to 30 cm long, 8 cm wide (**T** 8, Rondo
　　Plateau) ...................................... 8. *G. lutambensis*
　　Axes of inflorescence slender and less erect without such
　　distinct bracts at nodes; leaves less coriaceous, up to 20
　　cm long, 7 cm wide but nearly always much smaller (**K** 7,
　　**T** 3, 6, 7) ...................................... 5. *G. scheffleri*

1. **Gomphia reticulata** *Beauv.*, Fl. Oware 2: 22, t. 72 (1810); Oliver in F.T.A.: 320 (1868). Type: Nigeria, Oware, *Palisot de Beauvois* (G, holo.)

Shrub or small tree 1–12 m tall with fissured bark and rose-coloured wood. Leaves mostly thin and papery but sometimes thicker; lamina narrowly elliptic to elliptic, oblong-elliptic or slightly oblanceolate, (4.2–)7–15(–23.5) cm long, (1.7–)2–5(–6) cm wide, ± acuminate to usually very narrow acute apex, narrowly cuneate at the base, finely and regularly serrate or subentire to quite entire, sometimes crinkled; venation with (6–)8–17 pairs of secondary veins and very numerous close transverse tertiary veins at right angles to the midrib but almost no reticulate tertiaries, prominent on both sides; petioles 2–6 mm long; stipules triangular, 1–6 mm long, acute, usually deciduous or persistent on young growth. Raceme-like inflorescences 2–3, usually with very slender rhachis, grouped in terminal panicles, 5–20(–30) cm long, the cymules 1–7-flowered (1–2 in East Africa) usually with a persistent bract at the base; pedicels 5–12(–20) mm long, jointed at 1–6 mm from the base; bracts linear-triangular, 1–3 mm long, finely pointed, mostly soon deciduous. Sepals ovate, 4–9 mm long, 2–3 mm wide, slightly accrescent and red in fruit. Petals orange-yellow, broadly obovate, 5–12 mm long, 3–10 mm wide, slightly emarginate at the apex, cuneate at the base. Anthers 3–10 mm long. Drupelets black, ellipsoid, 6–9 mm long, 4–5 mm wide. Seeds with embryo heterocotylous, the small cotyledon external, incumbent.

UGANDA. Kigezi District: Ishasha Gorge, 27 Feb. 1998, *Hafashimana* 479!
KENYA. Kwale District: Shimba Hills, Longo Mwangandi, 16 Apr. 1968, *Magogo & Glover* 898! & 18 Mar. 1991, *Luke & Robertson* 2736! & Mwele Forest, 10 Oct. 2001, *Robertson* 7392!
TANZANIA. Lushoto District: East Usambara Mts, Kwamkoro Forest Reserve, 18 Nov. 1986, *Ruffo & Mmari* 1997!; Morogoro District: Nguru Mts, Manyangu Forest Reserve, 18 Sept. 1960, *Paulo* 796!; Ulanga District: Sanje loggers camp, *Rodgers* 386!
DISTR. **U** 2; **K** 7; **T** 3, 6, 7; W Africa, Congo (Kinshasa), Bioko
HAB. Evergreen forest including rain-forest, riverine forest; 300–1500 m

SYN. *Ouratea reticulata* (Beauv.) Gilg in E & P., Pf. ed. 1, 3 (6): 142 (1893); Engl. in E.J. 17: 79 (1893); Vollesen in Opera Bot. 59: 25 (1980)
　　*O. warneckei* Engl., V.E. 3 (2): 490 (1921) in clav.; Gilg in E & P. Pf. ed. 2, 21: 74 (1925); T.T.C.L.: 386 (1949). Type: Tanzania, Lushoto District: Amani, *Warnecke* 468* (B†, holo., EA!, iso.)
　　*Campylospermum reticulatum* (P. Beauv.) Farron in B.J.B.B. 35: 400 (1965) & in F.C.B., Ochnaceae: 51 (1967)

NOTE. In his B.J.B.B. paper (1965) Farron indicates from his synonymy that *C. reticulatum* is widely distributed in West to East Africa but in the F.C.B. account no distribution or synonymy outside the Congo is mentioned. Whether this is an accidental omission or due to some subsequent doubt over synonymy is not known. I am content to accept Farron's synonymy given in 1965. *Iversen* 87/213 is described as a 'scandent tree'. The single Uganda specimen seen has inflorescences only 7–8 mm long with two side branches, more graceful than in coastal material, but can be matched by much material from Congo (Kinshasa) and Cameroon.

* This specimen was not mentioned by Engler or Gilg but is clearly the type.

© The Board of Trustees of the Royal Botanic Gardens, Kew, 2005

Subsp. ?

Flowers up to 23 mm wide, the petals up to 11 mm long, 8 mm wide. Sepals 7 mm long, 3 mm wide. Leaves with entire margins.

TANZANIA. Lindi District: southern face of Rondo escarpment, Mchinjiri, Dec. 1951, *Eggeling* 6425!
DISTR. **T** 8; not known elsewhere
HAB. Evergreen forest; ± 690 m

SYN. *Ouratea* 'as yet unnamed species', Robson in F.Z. 2: 258 (1963) adnot.

NOTE. This is quite distinctive but only one gathering appears to be known. Farron identified the Kew sheet as *Campylospermum reticulatum* (P. Beauv.) Farron var. *reticulatum*. His maximum measurements for petals in var. *reticulatum* and var. *turnerae* (Hook. f.) Farron are as large and I have seen very similar large flowered specimens from Sierra Leone. Nevertheless it is distinctive amongst the East African material and more material may reveal subspecific characters. Farron separates var. *turnerae* on whether the leaves dry green or brown and *Iversen et al.* 87/213 from Kwamkoro Forest Reserve certainly has brown leaves but other material from the same place has them green or only slightly brownish. The relationship in West Africa between the varieties is very confused.

2. **Gomphia likimiensis** (*De Wild.*) *Verdc.* **comb. nov.** Type: Congo (Kinshasa), Likimi, *Malchair* 271 (BR!, syn.)

Shrub or small tree 1.2–4.5 m tall with smooth grey or brown bark; slash pale pink. Leaves thin, narrowly oblanceolate to narrowly elliptic, 4–21 cm long, 0.8–4.8 cm wide, narrowly acute to acuminate at the apex, narrowly cuneate at the base, closely serrate with curved aculeate teeth; midrib prominent on both surfaces and lateral veins ± 12–15, strongly curved upwards with numerous secondary ± parallel veins ± at right angles to the midrib and reticulate tertiary venation between, all prominent on both sides; petiole 1–4 mm long; stipules linear, 5–10 mm long, aristate, very soon deciduous. Inflorescences subterminal, 7–18 cm long, branched, the cymes 1–5-flowered; pedicels 6–10 mm long with joint 1–5 mm from base; bracts ovate, striate, deciduous. Sepals narrowly oblong-lanceolate, 5–8 mm long, 1.5–2.2 mm wide. Petals yellow, obovate, (5–)8 mm long, (2.5–)4.5–5.5 mm wide. Drupelets 1–3, reddish, presumably turning black, 5–6 mm long, 3–5 mm wide.

UGANDA. Bunyoro District: Budongo Forest, May 1932, *Harris* 92 in F.D. 725!; Mengo District: Kiagwe, Namanve Forest, May 1932, *Eggeling* 432 in F.D. 707! & Mulange, Mabira Forest, Oct. 1922, *Dummer* 5597!
KENYA. North Kavirondo District: Kakamega, Yala R., Apr. 1934, *Dale* in F.D. 3244!
TANZANIA. Mwanza District: Geita, Mwanza area, Buhindi Forest Reserve, 27 Aug. 1964, *Carmichael* 1080! & prob. Ukerewe I., 1929, *Conrads* 5960!
DISTR. **U** 2, 4; **K** 5; **T** 1; Congo (Kinshasa), Sudan
HAB. Forest edges and shrub layer, sometimes in open places; 1050–1500 m

SYN. *Ouratea likimiensis* De Wild. in Rev. Zool. Afr. 7, Suppl. Bot.: B59 (1920)
　　*O. floribunda* De Wild. in Rev. Zool. Afr. 7, Suppl. Bot.: B51 (1920). Type: Congo (Kinshasa), Eala, *Pynaert* 522 (BR, holo., K!, iso.), *non* (St. Hil.) Engl. (1876) *nom. illegit.*
　　*O. bukobensis* sensu T.T.C.L.: 385 (1949) pro parte; I.T.U. ed. 2: 281 (1951) pro parte; K.T.S.: 341 (1961) pro parte
　　*Campylospermum bukobense* sensu Farron in F.C.B. Ochnaceae: 50 (1967) pro parte, *non* (Gilg) Farron sensu stricto
　　*Ouratea hiernii* sensu Beentje, K.T.S.L.: 124 (1994), *non* (Tiegh.) Exell

NOTE. The taxon described above was for long called *Ouratea floribunda* De Wild. and the type of that exactly fits the Uganda populations. Unfortunately the name cannot be used in *Gomphia* since there is a prior *G. floribunda* St. Hil. See note at end of *G. vogelii*.

3. **Gomphia lunzuensis** (*N. Robson*) *Verdc.* **comb. et stat. nov.** Type: Zambia, Lunzua R., 30 km W of Mbala [Abercorn], above the falls, *Bullock* 3877 (K!, holo.)

© The Board of Trustees of the Royal Botanic Gardens, Kew, 2005

Shrub or small tree 1.5–6 m tall with pale brown slender wiry branches. Leaves subcoriaceous; lamina elliptic to narrowly oblong, 4–17.5 cm long, 1.3–4.3 cm wide, sharply acute to slightly acuminate at the apex, narrowly cuneate at the base, spinulose-serrate to subentire and usually quite entire near base; lateral veins 8–16, ascending, with numerous secondary veins almost at right angles to the midrib and reticulate tertiary veins, all prominent on both sides; petiole 2–5 mm long; stipules linear-triangular, 2.5–3.5 mm long, striate, very acute, persistent on young growth. Flowers 1–2(–3) in axils of bracts (which are persistent at least in anthesis) along a terminal or axillary very condensed almost spherical panicle 5–6 cm long and wide, shorter than the leaves; rhachis slender, green; basal bracts linear-triangular, 2.5 mm long, finely pointed, persistent; peduncle 0–1.5 cm long; pedicels 8–22 mm long in fruit, articulated 1.5–5 mm from the base. Sepals oblong, 5–7 mm long, 1.5–2 mm wide, rounded at the apex, becoming red and up to 9 mm long, 3 mm wide in fruit. Petals bright yellow or orange-yellow, narrowly obovate, 8–10 mm long, 4–5 mm wide, rounded or retuse at the apex. Anthers yellow, 4.5–6 mm long. Drupelets at first red, eventually black, cylindric-ellipsoid, 7–8 mm long, 5–5.5 mm wide.

TANZANIA. Buha/Kigoma District: Gombe National Park, Mkenke Valley, 20 Aug. 1969, *Clutton-Brock* 253!; Mpanda District: Kungwe-Mahali peninsula, between Pasagulu and Musenabantu, 8 Aug. 1959, *Harley* 9221! & 40 km N of Mpanda, Nyamansi R., Sept. 1961, *Procter* 1947!
DISTR. **T** 4; Zambia
HAB. Riverine and dry gallery forest; 900–1500 m

SYN. *Ouratea lunzuensis* N.Robson in Bol. Soc. Brot., Sér. 2, 36: 38 (1962) & in F.Z. 2: 258, t. 47a (1963)
    *Campylospermum reticulatum* sensu Farron in B.J.B.B. 35: 401 (1965) pro parte quoad syn. *O. lunzuensis*

NOTE. I do not agree with sinking Robson's species into *G. reticulata*. It has a quite different inflorescence structure and a distinct facies and occupies a circumscribed distribution in an area known for endemics.

4. **Gomphia vogelii** *Hook. f.*, Nig. Fl.: 272 (1849); Oliv. in F.T.A. 1: 321 (1868) pro parte. Type: Liberia, Grand Bassa, *Vogel* 53 (K!, holo.)

Shrub or small tree 1.5–12 m tall. Leaves fairly thin to distinctly subcoriaceous, mostly oblong or oblong-elliptic, less often elliptic-oblanceolate or narrowly elliptic, 3.5–15.5 cm long, 1–6.4 cm wide, acute or acuminate at the apex, narrowly cuneate at the base, serrate or often almost entire; lateral veins ± 9, ascending, in depressions above but actual vein prominent, more prominent beneath; secondary veins numerous, ± parallel and at right-angles to midrib and tertiary reticulate venation between; petiole ± 4(–6) mm long; stipules 3 mm long, striate, aristate, soon deciduous. Inflorescences subterminal, 4–12 cm long, shorter or sometimes much shorter than subtending leaf; cymes 1–2(–6)-flowered; peduncle ± 1 cm long; pedicels 5–6 mm long, enlarging to 13 mm in fruit. Sepals 6–8 mm long, 2.5–3 mm wide. Petals bright yellow, 5–8 mm long, 2.5–4 mm wide. Drupelets black, ellipsoid, 7 mm long, 5 mm wide.

UGANDA. Masaka District: Sese Is., Buire, June 1925, *Maitland* 796!* & Nkose I., 21 Jan. 1956, *Dawkins* 855! & Malabigambo Forest, 64 km SSW of Katera, 2 Oct. 1953, *Drummond & Hemsley* 4544!
TANZANIA. Bukoba District: Rubare Forest Reserve, Oct. 1957, *Procter* 711! and same locality, Nov. 1958, *Procter* 1057! & Ruasina Forest [Rwasina], Nov. 1935, *Wigg* 317! & Kaigi, May 1935, *Gillman* 275!

* There is confusion concerning *Maitland* 796. Two sheets 1 and 2 clearly state Sese I, Buire, June 1925, 3800' and one is annotated *Campylospermum vogelii* var. *poggei* by Farron; a third sheet also 796 has Ruwenzori Region at Buire, Dec. 1926, 4000' and a note 'gathered from the ? of the forest on the Semiliki side of the Ruwensori Mts'.

© The Board of Trustees of the Royal Botanic Gardens, Kew, 2005

DISTR. **U** 2, 4; **T** 1; W Africa, Congo (Kinshasa)
HAB. Swamp forest, secondary evergreen forest; 1100–1200 m

SYN. *Ouratea vogelii* (Hook. f.) Gilg in E & P. Pf. 3 (6): 141 (1893) & in E.J. 33: 248 (1903);
F.W.T.A. 1: 229 (1954)
O. *bukobensis* Gilg in E.J. 33: 271 (1903); Z.A.E. 2: 559 (1913); T.T.C.L.: 385 (1949) pro
parte; I.T.U. ed. 2: 281 (1951) pro parte. Type: Tanzania, Bukoba District: near Bukoba
on Lake Victoria, *Stuhlmann* 978, 1023, 1073, 1467, 1521, 3212, 3721, 3990a, 3991, 4063
(all B†, syn.)*
*Campylospermum bukobense* (Gilg) Farron in B.J.B.B. 35: 393 (1965) & in F.C.B. Ochnaceae:
50 (1967) excl. syn. *Ouratea floribunda* De Wild.
C. *vogelii* (Hook. f.) Farron in B.J.B.B. 35: 404 (1965)

NOTE. Material referred to *G. bukobensis* in East Africa during the past century can be more or
less sorted into two taxa. One has thin narrowly elliptic-lanceolate leaves with closer more
curved usually spinulose teeth. Much of this has been determined as *Ouratea floribunda* De
Wild. and matches the type of that and many specimens of it are determined as
*Campylospermum bukobense* by Farron. The other has more coriaceous more oblong, often
wider leaves with fewer less spinulose teeth or sometimes partly entire and at least one
specimen of this has been annotated as *C. vogelii* var. *poggei* (Engl.) Farron by Farron. I have
not seen any type material of *Ouratea bukobensis* but many specimens are available from the
Bukoba area; none of these at Kew have been annotated by Farron but I believe they belong
to the second taxon. Robson has annotated Bukoba material as *Ouratea hiernii* (Tiegh.) Exell
in which he includes *O. bukobensis* and *O. poggei* (Engl.) Gilg but excludes *O. floribunda* De
Wild (see F.Z. 2: 256–7 (1963)) thus agreeing with my treatment. It must be admitted that the
differences are small and specimens often difficult to place, the serration being particularly
variable. *Ouratea bukobensis* and *O. floribunda* have been considered synonymous by most
workers and Farron has continued this but he annotated a specimen of true *bukobensis* from
Uganda as *vogelii*. I am keeping up two species and hope that some really informative field
work will settle the problem.

Farron records *Campylospermum vogelii* var. *poggei* from Sudan, Congo (Kinshasa), Rwanda,
Burundi, Uganda, Kenya, Tanzania and Angola; he also gives the maximum height attained
as 25 m and trunk 45 cm in diameter. Nothing of this size has been found in East Africa.
Nowhere does he say how the variety differs from the species.

5. **Gomphia scheffleri** (*Gilg*) *Verdc.* **comb. et stat. nov.** Type: Tanzania, Lushoto
District: East Usambaras, Derema, Ugamba Forest, *Scheffler* 189 (B†, holo., EA!, iso.,
K!, iso.)

Shrub or small tree 1.5–15(–20) m tall with grey-brown bark and slash orange, red
beneath. Leaves ± subcoriaceous, elliptic to oblong-elliptic or oblong, 4–20 cm long,
2–7 cm wide, narrowed to a bluntly acute or obtuse apex or more rounded, cuneate
or ± rounded at the base, distinctly or shallowly and inconspicuously serrulate or ±
subentire; lateral veins 13–20 each side, the ultimate venation either with very
numerous and very close parallel veinlets or with few of these or more openly
reticulate with scarcely any; petiole 5–10 mm long; stipules narrowly triangular, 3 mm
long, striate. Inflorescences lax, the axes subsimple and racemiform, 3–15 cm long
with peduncle etc. flattened; pedicels 6–11 mm long, jointed at base. Sepals elliptic,
3.5–5.5 mm long, 1.5–4 mm wide. Petals bright yellow, oblanceolate-oblong to
broadly obovate, 4.5–10 mm long, 1.5–6 mm wide, obtuse to broadly rounded or
retuse at the apex. Anthers orange, 2.5–5 mm long. Drupelets red, (1–)2(–5?),
obovoid, 7 mm long, 6 mm wide (probably black when ripe).

1. Petals smaller and narrower, 4.5–5 × 1.5–2 mm; ultimate
   veinlets very numerous, very close and parallel (**T** 3, East
   Usambaras) . . . . . . . . . . . . . . . . . . . . . . . . . . . . . . . . .  a. subsp. *scheffleri*
   Petals larger; ultimate veinlets more openly reticulate or
   close ones less numerous . . . . . . . . . . . . . . . . . . . . . . . . . . . . . . . . . . . . 2

* I think at least one of these syntypes, 1467, came from Uganda, Manjonjo (=Munyonyo).

© The Board of Trustees of the Royal Botanic Gardens, Kew, 2005

2. Petals 9 × 5 mm; ultimate veinlets more openly reticulate
   giving a very different appearance (**K** 7, Teita); **T** 3, West
   Usambaras) ................................... b. subsp. *taitensis*
   Petals 6.5–10 × 3.5–6 mm; ultimate veinlets more similar to
   subsp. *scheffleri* but less numerous and somewhat
   intermediate (**T** 6, 7) ........................... c. subsp. *schusteri*

a. subsp. **scheffleri**

Shrub 1.5–4 m or less often a small shrubby tree to 8 m or tree to 20 m. Leaves more elliptic and with a very distinct ultimate venation of numerous very close parallel veinlets; petals narrowly oblong-oblanceolate, 4.5–5 × 1.5–2 mm.

TANZANIA. Lushoto District: East Usambaras, Amani, Apr. 1926, *B.D. Burtt* 446! & Sangarawe, 21 Nov. 1947, *Brenan & Greenway* 8359! & Amani, 18 Apr. 1922, *Don Carlos* in AH 6049!
DISTR. **T** 3 (East Usambaras); not known elsewhere
HAB. Rain-forest; 1000–1450 m

SYN. *Ouratea scheffleri* Gilg in E.J. 33: 262 (1903)

NOTE. *Iversen* et al. 87/468 is described as a "scandent tree" to 8 m and *Grote* AH 3886 is described as "20 m glatten stamm".

b. subsp. **taitensis** Verdc. **subsp. nov.** a subsp. *scheffleri* foliis oblongioribus, rete venularum tertiariarum minus crebro, petalis obovatis 9 × 5 mm, late rotundatis vel retusis differt. Type: Kenya, Teita District: Taita Hills, 8 km NNE of Ngerenyi, Ngangao, *Drummond & Hemsley* 4342 (K!, holo., iso., EA, iso.)

Usually a small tree 9–10 m tall. Leaves more oblong with more open reticulate venation. Petals obovate, ± 9 × 5 mm, broadly rounded to retuse at apex.

KENYA. Teita District: Teita Hills, 8 km NNE of Ngerenyi, Ngangao, 15 Sept. 1953, *Drummond & Hemsley* 4342! & Mbololo Forest, 13 May 1985, *Faden et al.* THE 407! & Kasigau Mt, Rukanga route, 1 June 1998, *Luke et al.* 5364!
TANZANIA. Lushoto District: West Usambaras, Shagayu [Shagai] Forest Reserve, June 1951, *Eggeling* 6152! & same locality, May 1953, *Procter* 217! & same locality, 2 km SE of Sunga, 2 Mar. 1953, *Drummond & Hemsley* 1414!
DISTR. **K** 7; **T** 3 (West Usambaras); not known elsewhere
HAB. Evergreen forest including mist forest, *Ocotea – Podocarpus* etc.; 1450–1950 m

SYN. *Ouratea* sp. nov.; K.T.S.: 343 (1961)
   *O. schusteri* sensu Beentje in Utafiti 1 (2): 50 (1988) pro parte & in K.T.S.L.: 124 (1994)

c. subsp. **schusteri** (Engl.) Verdc. **comb. et stat. nov.** Type: Tanzania, Morogoro District: Uluguru Mts, Lupanga Peak, no collector or specimens cited (B†, holo.)

Mostly an evergreen shrub 4.5–5 m tall or tree to 15 m (**T** 7). Leaves oblong or elliptic-oblong with final venation with close parallel veinlets but not so numerous and marked as in subsp. *scheffleri*. Petals obovate, 6.5–10 × 3.5–6 mm.

TANZANIA. Morogoro District: Lupanga Peak to Morogoro track, 16 Aug. 1951, *Greenway and Eggeling* 8620! & Lupanga Peak, 26 June 1983, *Polhill & Lovett* 4921!; Iringa District: above Sanje village, Mwanihana Forest Reserve, 10 Oct. 1984, *D.W. Thomas* 3834!
DISTR. **T** 6, 7; not known elsewhere
HAB. Dry evergreen forest, ridge forest; 1600–2050 m

SYN. *O. schusteri*\*Engl. in V.E. 3 (2): 489 (1921); Gilg in E & P., Pf. 21: 73 (1925); T.T.C.L.: 586 (1949); Beentje in Utafiti 1 (2): 50 (1988) pro parte

NOTE. *Lovett* 144 collected in the type locality is described as a liane.

\* Presumably named after Curt Schuster, librarian at the Berlin Botanic Garden, and not after a collector.

© The Board of Trustees of the Royal Botanic Gardens, Kew, 2005

6. **Gomphia sacleuxii** (*Tiegh.*) *Verdc.* **comb. et stat. nov.** Type: "Zanzibar Mbonda" i.e. Tanzania, Morogoro District: Mhonda, *Sacleux* s.n. & Nguru, *Sacleux* 1551 (both P, syn.)

Shrub or small tree 1–5(–6) m tall, sometimes with drooping branches. Leaves ± coriaceous, oblong-elliptic to elliptic or broadly oblanceolate, 15–33 cm long, 4–12 cm wide, shortly acuminate, subacute or acute (± obtuse in some damaged leaves) at the apex, cuneate to rounded or subcordate (or rarely very narrowed) at the base, serrate, the teeth ± black-tipped when dry, or in places subentire; lateral veins 13–20, lying in depressed hollow above but actual veins prominent, very prominent beneath; numerous veins between the veins and tertiary venation of very numerous parallel veinlets, between the secondaries, more or less in direction of laterals, prominent; petiole 4 mm long; stipules lanceolate, 3–7 mm long, striate, also present on internodes and numerous in a short distance without any associated leaves. Inflorescences pseudo-racemose, up to 35 cm long, with 1–6-flowered cymes from node-like abbreviated branches; peduncles 10–12 cm long; pedicels 10–15 mm long, jointed 1–3 mm above the base. Sepals narrowly elliptic-ovate, 8 mm long, 3–3.5 mm wide, becoming orange-red in fruit. Petals yellow or orange-yellow, elliptic, 8–12 mm long, 4–6 mm wide, rounded at apex. Anthers narrowed to apex, 5–6 mm long. Drupelets scarlet, subglobose, 6 mm long, 5–6 mm wide (probably eventually turn black). Fig. 7 (p. 51).

KENYA. Kwale District: Shimba Hills Game Reserve about 1.6 km E of junction with main road, on road to Giriama point, 7 Dec. 1972, *Spjut & Ensor* 2728! & Longomwagandi Forest, 21 Feb. 1968, *Magogo & Glover* 166! & Gongoni Forest, 28 Nov. 1996, *Luke* 4542!
TANZANIA. Tanga District: Sigi Valley, 13 km below Amani, 29 Dec. 1956, *Verdcourt* 1750!; Morogoro District: between Turiani Falls and Mhonda sawmill, 4 Nov. 1947, *Brenan & Greenway* 8284!; Iringa District: Sanje Falls, 23 July 1983, *Polhill & Lovett* 5122!
DISTR. **K** 7; **T** 3, 6, 7; not known elsewhere
HAB. Lowland evergreen forest including riverine forest, ground water forest and semi-dry forest; 40–750 m

SYN. *Cercanthemum sacleuxii* Tiegh. in Ann. Sci. Nat. ser. 8 Bot. 16: 308 (1902)
　　*Ouratea* sp.; T.T.C.L.: 386 (1949) quoad *Burtt* 5406
　　*Campylospermum sacleuxii* (Tiegh.) Farron in B.J.B.B. 35: 401 (1965)
　　*Ouratea sacleuxii* (Tiegh.) Beentje in Utafiti 1 (2): 70 (1988); K.T.S.L.: 124, fig. (1994)

7. **Gomphia densiflora** (*De Wild. & Dur.*) *Verdc.* **comb. et stat. nov.** Type: Congo (Kinshasa), without locality, *Dewèvre* s.n. (BR, holo.)

Shrub or small tree, sometimes straggling, (1–)2–5(–12) m tall with small crown; bark dark grey, ± smooth save for lenticels or flaking; slash crimson red. Leaves coriaceous, ± shiny on both surfaces, oblong-obovate to elliptic, (10–)12–32(–37) cm long, (4–)6–12.5 cm wide, acute to subacuminate or obtuse to rounded at the apex, cuneate, rounded or truncate at the base, serrulate at least towards base or often almost entire, the venation prominent on both surfaces but mostly faint especially above; petiole 3–10(–12) mm long, 3–4 mm wide; stipules narrowly triangular, 2–6.5 mm long, 2 mm wide, scarious, usually deciduous. Flowers in solitary raceme-like inflorescences or in extensive terminal panicles, 10–30 cm long, the cymules 1–11-flowered; peduncle and rhachis ± angular; bracts triangular, 5–6 mm long, acute, soon deciduous; pedicels 7–15 mm long, articulated 2–4 mm from the base. Sepals elliptic, 5–7 mm long, 2–3.5 mm wide, persistent, the fruiting sepals accrescent, red, 8–9 mm long, 4 mm wide; petals yellow or orange, obovate, 6–10 mm long, 4–7 mm wide, unguiculate, deciduous; anthers orange yellow, 3–5 mm long. Drupelets black, ellipsoid or subglobose, 7–10 mm long 4–6 mm wide. Seeds with embryo heterocotylous, the small cotyledon external.

© The Board of Trustees of the Royal Botanic Gardens, Kew, 2005

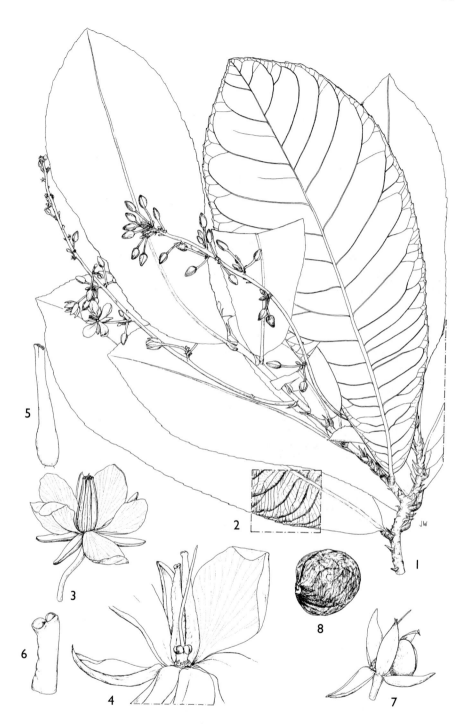

FIG. 7. *GOMPHIA SACLEUXII* — **1**, habit, × ¹/₂; **2**, detail of upper leaf surface, × ¹/₂; **3**, flower, × 1 ¹/₂; **4**, flower details, × 4; **5**, anther, × 6; **6**, anther detail, × 16; **7**, young fruit, × 2; **8**, drupelet, × 3. 1–2 from *Brenan* 8284, 3 from *Kisena* 1442, 4, 6 from *Lovett* 217, 5 from *Magogo & Glover* 166, 7–8 from *Verdcourt* 1750. Drawn by Juliet Williamson.

© The Board of Trustees of the Royal Botanic Gardens, Kew, 2005

UGANDA. Bunyoro District: Siba Forest, *Sangster* 112!; Masaka District: 6.4 km SSW of Katera, Malabigambo Forest, 2 Oct. 1953, *Drummond & Hemsley* 4568!; Mengo District: Kyagwe, near Mukono, Kasara R., 20 Dec. 1950, *Dawkins* 684!

KENYA. North Kavirondo District: Kakamega, ?July 1944, *Carroll* H62/44!

TANZANIA. Bukoba District: Kikuru Forest, Sept./Oct. 1935, *Gillman* 446! & Minziro Forest, Aug. 1957, *Procter* 650!; Mpanda District: Mahali Mountains National Park, 5 Sept. 1991, *Hamai* 91004!

DISTR. U 2, 4; **K** 5; **T** 1, 4; Nigeria, Cameroon, Equatorial Guinea, Gabon, Central African Republic, Congo (Kinshasa), Sudan and Zambia

HAB. Evergreen and semi-deciduous forest including swamp forest; 800–1500 m

SYN. *Ouratea densiflora* De Wild. & Dur. in Ann. Mus. Congo Belge Bot. ser. III, 1: 37 (1901); Gilg in E.J. 33: 265 (1903) & in Z.A.E.: 558 (1913); F.P.N.A. 1: 616, t. 61 (1948); T.T.C.L.: 385 (1949); I.T.U. ed. 2: 281 (1951); F.P.S. 1: 188 (1950); K.T.S.: 341 (1961); White, F.F.N.R.: 252, fig. 44 (1962); Robson in F.Z. 2: 256 (1963); K.T.S.L.: 124 (1994)

    *O. coriacea* De Wild. & Dur. in Ann. Mus. Congo Belge Bot. Ser. III, 1: 36 (1901); Gilg in E.J. 33: 261 (1903); F.P.N.A. 1: 616 (1948). Type: Congo (Kinshasa), between Lukolela and Ngombi, *Dewèvre* 795 (BR, holo.)

    *Campylospermum densiflorum* (De Wild. & Dur.) Farron in B.J.B.B. 35: 394 (1965) & in F.C.B. Ochnaceae: 46, t. 5 (1967)

NOTE. Although there is an entry *G. densiflora* Spruce ex Engl. given in Index Kewensis this does not invalidate De Wildeman and Durand's name when their use of the same epithet is transferred to *Gomphia* because Engler merely mentions Spruce's name in synonymy under *Ouratea acuminata*.

### 8. **Gomphia lutambensis** (*Sleumer*) *Verdc.* **comb. et stat. nov.** Type: Tanzania, Lindi District, Lake Lutamba, Noto Plateau, *Schlieben* 6110 (B†, holo.)

Shrub 4–5 m tall (scandent fide Bidgood et al.); bark brown, shining, longitudinally striate, splitting. Leaves shining in life, less so and brownish on drying, oblong, elongate-oblong or elliptic-oblong, 10–30 cm long, 3–8 cm wide, rounded at apex and base, margin crenulate, entire or with few very obscure traces of serration; midrib ± impressed above, very prominent beneath; lateral veins ± 35 with venation very densely reticulate and with some close parallel veinlets, prominent on both sides; petiole thick, 3–5 mm long; stipules triangular, 5 mm long, striate, thickened. Panicles terminal with very abbreviated branches bearing numerous persistent bracteoles, with only terminal flowers well developed; rhachis 6–13 cm long; pedicels 1.2–1.5 cm long. Sepals thin, ovate-oblong, 6–7 mm long, 3 mm wide. Petals yellow, oblong-spathulate, 9–10 mm long, 5 mm wide. Stamens 10; anthers ± 7 mm long, almost sessile. Drupelets black, subglobose, ± 6 mm in diameter (immature).

TANZANIA. Lindi District: Noto Plateau, Lake Lutamba, [15 July 1935]*, *Schlieben* 6110 & Rondo Forest Reserve, 15 Feb. 1991, *Bidgood et al.* 1581! & S face of Rondo escarpment, Mchinjiri, Dec. 1951, *Eggeling* 6416!

DISTR. **T** 8; not known elsewhere

HAB. Semi-evergreen forest with *Milicia, Albizia, Dialium, Pteleopsis* on grey sandy soil; 450–750 m

SYN. *Ouratea lutambensis* Sleumer in F.R. 39: 279 (1936); T.T.C.L.: 385 (1949)

NOTE. Sleumer says this is related to *G. calophylla* Hook. f. (*Ouratea calophylla* (Hook. f.) Engl., *Rhabdophyllum calophyllum* (Hook. f.) Farron) but this is not so since the latter has much closer lateral veins. I do not think Farron would have referred Sleumer's species to *Rhabdophyllum*. *Eggeling* 6116 is undoubtedly identical with *Bidgood et al.* 1581 and both had been identified as *lutambensis* by the collectors. Farron confirmed the *Eggeling* sheet as *lutambensis* in 1963. Robson had annotated it 'not *lutambensis* which belongs to Calophyllae'. It is not clear if Farron or Robson saw type material.

* Date from Leistner in Bothalia 12: 133 (1976).

© The Board of Trustees of the Royal Botanic Gardens, Kew, 2005

9. **Gomphia mildbraedii** (*Gilg*) *Verdc.* **comb. et stat. nov.** Types: Congo (Kinshasa), Irumu-Mawambi, *Mildbraed* 2918, 2931 (both B†, syn.)

Shrub to 4 m or small tree to 6 m. Leaves ± thin, lamina elliptic or obovate-elliptic, 5.5–13(–15) cm long, 2.5–5.5 cm wide, abruptly acuminate at the apex, narrowly cuneate at the base, entire or rarely some leaves with 1–3 teeth with a distinct seta directed apically; venation itself slightly prominent on both surfaces but above the lateral veins lie in slightly depressed furrows; petiole 4–7 mm long; stipules very narrowly triangular, 4 mm long, striate. Inflorescences axillary, short, sessile, 1–3-flowered; pedicels 10–14 mm long, articulated at the base. Sepals in bud 5–7 mm long, 2–3 mm wide, enlarging to 15 mm long and 5 mm wide in fruit and becoming red. Petals yellow, elliptic, 4 mm long, 2–3 mm wide; anthers 4 mm long. Drupelets ellipsoid, 13 mm long, 6–7 mm wide.

UGANDA. Bunyoro District: Budongo Forest, Apr. 1941, *Eggeling* 4255!
DISTR. **U** 2; Congo (Kinshasa)
HAB. Evergreen forest; ± 1050 m

SYN. *Ouratea mildbraedii* Gilg in Z.A.E.: 558, t. 75 (1913)
   *O. morsoni* sensu Dale, I.T.U. ed. 2: 282 (1952), *non* Hutch. & Dalz.
   *Idertia mildbraedii* (Gilg) Farron in Bull. Soc. Bot. Suisse 73: 212 (1963) & in F.C.B., Ochnaceae: 23 (1967)

## 4. LOPHIRA

Gaertn. in Gaertn. f., Fruct. 3: 52, t. 188, fig. 2 (1805); Bamps in B.J.B.B. 40: 291–294 (1970)

Trees; branches with very obvious leaf-scars. Leaves tufted at tips of branches, simple, petiolate, entire; secondary veins numerous; stipules entire. Flowers in terminal panicles. Calyx lobes 5, imbricate in bud, persistent. Corolla contorted in bud; petals 5; stamens numerous in 3–5 whorls; anthers dehiscing by 2 apical pores. Ovary sessile, of 2 carpels, 1-locular with basal placentation, the ovules in 2 rows, erect, anatropous; style scarcely distinct from ovary; stigmas 2, divergent. Achenes subwoody, surrounded by calyx with outer sepals accrescent and wing-like. Seeds without endosperm.

Genus of 2 species restricted to tropical Africa.

**Lophira lanceolata** *Keay* in K.B. 8: 488 (1954) & in F.W.T.A. ed. 2, 1: 231 (1954); Keay, Onochie & Stanfield, Niger. Trees 1: 231 (1954); Irvine, Woody Pl. Ghana: 91, fig. 28 (1966); Dale, I.T.U. corrig. & addenda: 8 (1956); Bamps in F.C.B. Ochnaceae: 56 (1967) & in B.J.B.B. 40: 293 (1970); Keay, Nigerian Trees: 73, fig. 30 (1989). Type: Guinea, Kebali, *Maclaud* 443 (P, holo., K, photo.!)

Coarsely branched tree up to 9–16 m tall with clean bole up to 7.5 m tall; bark corky, grey, very coarsely flaking in small pieces; flakes brick-red to yellow beneath, brittle; slash bright yellow cork layer above, crimson-red granular beneath. Leaves red when young, oblong-lanceolate, 11–45 cm long, 2–9 cm wide, rounded to retuse at the apex, cuneate and often asymmetrical at base, entire, glabrous; lateral veins very numerous together with midrib prominent on both surfaces; petiole 2–6 cm long; stipules linear-lanceolate, 3–5 mm long, 0.7 mm wide, deciduous. Panicles profuse, terminal, pyramidal, lax, 15–20 cm long; axes angular, striate, glabrous. Flowers white, scented; pedicels 1–1.5 cm long, jointed near apex, glabrous. Sepals unequal, glabrous; 2 outer ovate-acuminate, 7–8 mm long, 4–5 mm wide, acute at apex; inner 3 broadly ovate, 6 mm long, 5 mm wide, obtuse at apex. Petals obcordate,

© The Board of Trustees of the Royal Botanic Gardens, Kew, 2005

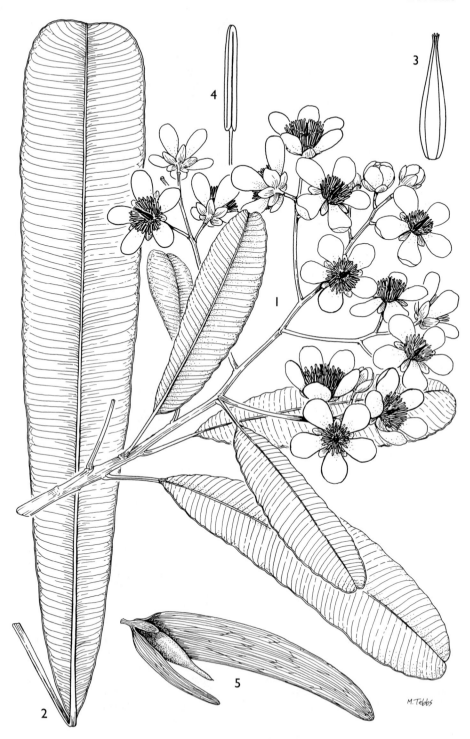

FIG. 8. *LOPHIRA LANCEOLATA* — **1**, flowering branch, × ²/₃; **2**, leaf, × ²/₃; **3**, ovary, × 3; **4**, stamen, × 6; **5**, fruit, × ²/₃. 1–3 from *Dawe* 1, 4 from *Schweinfurth* 84, 5 from *Schweinfurth* 1755. Drawn by Margaret Tebbs.

© The Board of Trustees of the Royal Botanic Gardens, Kew, 2005

1.7 cm long, 1.3 cm wide, glabrous. Stamens with filaments white, 4–6 mm long and anthers orange, 4–5 mm long. Ovary white, conical, 8 mm long, 3 mm wide; stigmas 1–2 mm long; ovules 8–16. Fruits conical, 3 cm long, 1 cm wide, glabrous with unequal wing-like enlarged sepals, one 8–10 cm long, 2–2.5 cm wide and the other 2.5–5 cm long, 0.6–1 cm wide. Seed 1, ovoid, 1.6 cm long, 8 mm wide. Fig. 8 (p. 54).

UGANDA. West Nile District: 200 yards N of where Amua leaves Otze Forest Reserve near boundary and Amua path, 5 Dec. 1947, *Dawkins* 300! & Otze Forest Reserve, 16 Dec. 1962, *Styles* 275!; Acholi District: Gulu, 15 May 1932, *Hancock in Tothill* 1118!
DISTR. **U** 1; Senegal to Cameroon, Central African Republic, Congo (Kinshasa) and Sudan
HAB. Open woodland with *Vitex*, *Hymenocardia*, *Butyrospermum*, *Combretum* etc. and annually burnt *Hyparrhenia – Oxytenanthera* undergrowth; 900–1500 m

SYN. *L. africana* G. Don in Gen. syst. 1: 814 (1831) pro parte quoad descr., *non* Banks ex G. Don
    *L. alata* sensu Oliv. in F.T.A. 1: 174 (1868) pro parte & in Trans. Linn. Soc. London 29: 33 (1873) (with extensive notes by J.A. Grant); F.W.T.A. ed. 1: 195 (1927) pro parte; Chalk and Burtt Davy, For. Trees & Timbers Brit. Emp. 2: 76 (1933) pro parte; Aubréville, Fl. For. Côte Iv. 2: 269 (1936) & Fl. For. Soud. Guin.: 80, t. 13 (1950); Dale & Eggeling, I.T.U. ed. 2: 276, fig. 61 (1952), *non* Gaertn. f.
    *L. lanceolata* Tiegh. in Journ. de Bot. 15: 187 (1901) *nom. provis.*
    *L. spatulata* Tiegh. in Journ. de Bot. 15: 187 (1901) *nom. provis.* pro parte quoad *Barter* 1167

NOTE. The wood is very tough and durable being stronger and harder than teak; the seeds yield oil.

## 5. SAUVAGESIA

L., Sp. Pl.: 203 (1753) & Gen. Pl. ed. 5: 95 (1754)

*Vausagesia* Baill. in Bull. Soc. Linn. Paris 2: 871 (1890); Robson in F.Z. 2: 262 (1963)

Completely glabrous shrubs, shrublets or annual or perennial herbs. Leaves sessile to petiolate, serrate; stipules with laciniate margins or dissected into filamentous lobes, persistent. Inflorescences paniculate terminal or few-flowered monochasial cymes or flowers solitary in axils of foliage leaves or reduced bract-like leaves. Sepals 5, quincuncially imbricate, persistent. Petals 5, white or pink, deciduous. Stamens 5, antisepalous, free with anthers dehiscing longitudinally; filaments short, in one species adnate to base of staminodial tube. Staminodes 5, petaloid, opposite the true petals in one species but usually in 2 whorls; outer staminodes many in a continuous ring or in 5 antisepalous groups, short, filiform and capitate or narrowly petaloid; inner staminodes 5, antipetalous, petaloid, free, forming a sort of corona. Ovary with 3 parietal placentas (or axile at the base) each with many ovules in 2 rows; style simple, slender with small stigma. Capsule with 3 septicidal valves; seeds numerous with punctate testa, abundant endosperm and straight embryo.

About 25 species all but 3 confined to tropical America. *Vausagesia* Baill. was kept as a monotypic genus until Bamps suggested that only one consistent floral character was not adequate to separate genera. Amaral's (E.J. 113: 105–196 (1991)) study of the family supports this and Robson (pers. comm.) thinks it is probably correct. Lebrun & Stork, Trop. Afr. Pl. 1: 506 (2003) still keep up *Vausagesia*. Only a detailed study of the American species will settle this problem.

Petaloid staminodes free, not united with the base of the staminal
  filaments; outer filiform staminodes free; flowers solitary or
  rarely paired in leaf axils . . . . . . . . . . . . . . . . . . . . . . . . . . . . . . . .    1. *S. erecta*
Petaloid staminodes connate at base, each united with the base of a
  staminal filament; outer filiform staminodes absent; flowers in
  monochasial cymes or rarely solitary in leaf axils . . . . . . . . . . .    2. *S. africana*

© The Board of Trustees of the Royal Botanic Gardens, Kew, 2005

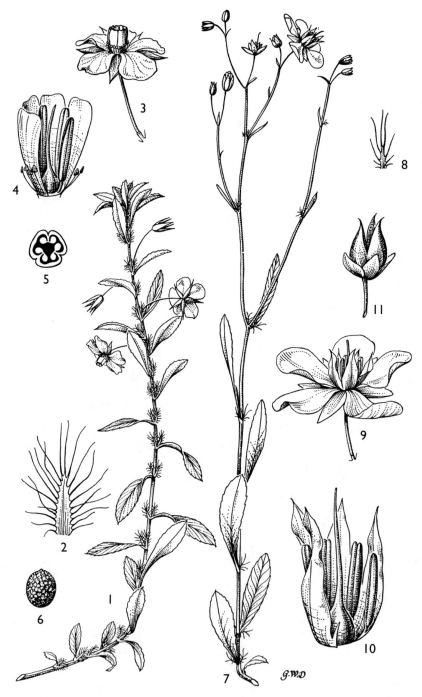

Fig. 9. *SAUVAGESIA ERECTA* — **1**, flowering shoot, × 1; **2**, stipule, × 4; **3**, flower, × 2; **4**, part of androecium and gynoecium, × 8; **5**, ovary transverse section, × 16; **6**, seed, × 20. *SAUVAGESIA AFRICANA* — **7**, flowering shoot, × 1; **8**, stipule, × 4; **9**, flower, × 2; **10**, part of androecium and gynoecium, × 8; **11**, dehisced capsule, × 2. 1–5 from *Fanshawe* 3630, 6 from *Walter* 28; 7–8 from *Milne-Redhead* 3030, 9–10 from *Marks* 56, 11 from *Gossweiler* 4110. Drawn by G.W. Dalby, and reproduced from Flora Zambesiaca.

© The Board of Trustees of the Royal Botanic Gardens, Kew, 2005

1. **Sauvagesia erecta** *L.*, Sp. Pl.: 203 (1753); Oliv., F.T.A. 1: 111 (1868); Gilg in E & P. Pf. ed. 2, 21: 83, fig. 43a (1925); F.P.S. 1: 188 (1950); Exell & Mendonça, C.F.A. 1: 284 (1951); Keay, F.W.T.A. ed. 2, 1 : 231 (1954); Perrier, Fl. Madag. 139, Violacées: 46 (1955); White, F.F.N.R.: 262 (1962); Robson in F.Z. 2: 260, t. 48, fig. A (1963); Bamps, F.C.B., Ochnaceae: 57 (1967); Vollesen in Opera Bot. 59: 26 (1980); U.K.W.F. ed. 2: 94 (1994). Type: Dominican Republic, Domingo, Linnaeus Herb. 283.2 (LINN, lecto., chosen by Robson)

Annual or perennial herb, semi-creeping or with erect shoots 5–60 cm tall, often with elongate branches from near the base, glabrous; stems red or purple-tinged, slender, wiry, angular. Leaves elliptic or oblong-elliptic to oblanceolate, 0.8–3 cm long, 2–9 mm wide, acute to obtuse or shortly apiculate at the apex, narrowed to the base, thickened and serrulate at the margin; petiole usually absent but sometimes up to 4 mm long, slender; stipules linear, 4.5–7 mm long, with long-fimbriate margins. Flowers axillary, solitary or rarely paired; pedicels purplish, (5–)8–20 mm long, very slender. Sepals elliptic, 3–7 mm long. Petals white to pink, obovate, 5–8 mm long, 3–5 mm wide, rounded at apex, spreading. Outer filiform staminodes numerous in an uninterrupted whorl, crimson or purplish, sometimes white above. Inner petaloid staminodes white, crimson to purple at the base, oblong-elliptic, 2.5–4.5 mm long, truncate to retuse at apex; anthers yellow, linear-oblong, 1.5–2 mm long. Ovary ovoid, 1 mm long; style slender, 1.5–2 mm long. Capsule ovoid, ± 5 mm long; seeds brownish orange, ellipoid or cylindric-ellipsoid, 0.5 mm long. Fig. 9/1–6 (p. 56).

UGANDA. Masaka District: Lake Nabugabo, NW side, 9 Oct. 1953, *Drummond & Hemsley* 4692! & Sese Is., Bugala I., Kagolomulo, 17 July 1951, *Norman* 19!; Mengo District: Kivuvu, May 1914, *Dummer* 879!
KENYA. Trans-Nzoia District: Kitale, 1942 (fide U.K.W.F.)
TANZANIA. Bukoba District: Bukoba, June 1931, *Haarer* 2017!; Kigoma District: about 4.8 km from Kigoma on Kasulu Road, 11 July 1960, *Verdcourt* 2788!; Rufiji District: Mafia I., Dawe Simba – Ndaagoni, 4 Oct. 1937, *Greenway* 5384!; Zanzibar I.: Kama Swamp, 4 Sept. 1963, *Faulkner* 3267!
DISTR. **U** 4; **K** 3; **T** 1, 4, 6; **Z**; **P**; widespread from Senegal to Sudan and S to Angola, Mozambique, Madagascar, also in W Indies and S America
HAB. Grassland in marshy areas, formations between grassland and *Papyrus* swamps, swamps on black cotton soils, swamp forest margins, sandy roadside gullies and irrigation channels, rice fields; 750–1200(–2100) m and at coast 0–9 m

2. **Sauvagesia africana** *(Baill.) Bamps* in F.A.C., Ochnaceae: 58 (1967). Type: ?Congo, neither locality nor collector cited (P, holo.)

Erect stems from rhizome 20–30(–45) cm tall, usually unbranched. Leaves linear to obovate or oblanceolate, 1.7–3.5 cm long, 0.2–1.2 cm wide, becoming progressively narrower up the stem, acute to obtuse or shortly apiculate at the apex, narrowed or cuneate at the base; petiole slender, up to 2 mm long or ± absent; stipules linear, 3–7 mm long, striate. Flowers in monochasial cymes or rarely solitary in leaf axils; pedicels 3–10 mm long, sometimes articulated at the base. Sepals lanceolate, 3–6 mm long. Petals pink with yellow and white base, obovate, 5–8 mm long, 3.5–4.5 mm wide, rounded. Staminodes white with pinkish-purple base, lanceolate, 3–5 mm long, rounded or slightly retuse at apex; anthers yellow, 1.5 mm long. Ovary ovoid, ± 1.5 mm long; style ± 2 mm long; stigma pink. Capsule ovoid; seeds ellipsoid, 0.5 mm long with punctate testa. Fig. 9/7–11 (p. 56)

UGANDA. Masaka District: Lake Nabugabo, Oct. 1932, *Eggeling* 581! & SW side of Lake Nabugabo, 7 Oct. 1953, *Drummond & Hemsley* 4670! & 2.5 km NE of Sand Beach Hotel, Lake Nabugabo, 15 Jan. 2002, *Lye & Namaganda* 25387!
DISTR. **U** 4; Congo (Kinshasa), Angola, Zambia

© The Board of Trustees of the Royal Botanic Gardens, Kew, 2005

Hab. In pure *Saccharum* stands, seasonally wet grasslands and *Sphagnum* bog; 1100–1200 m

Syn. *Vausagesia africana* Baill. in Bull. Soc. Linn. Paris 2: 871 (1890); Gilg in E & P. Pf. ed. 2,
       21: 84 (1925); Exell & Mendonça, C.F.A. 1 (2): 284 (1951); Robson in F.Z. 2: 262, t.
       48, fig. b (1963)
     *V. bellidifolia* Engl. & Gilg in Warb., Kunene-Samb. Exped.: 305 (1903); Gilg in E & P.
       Pf., ed. 2, 21: 84 (1925). Type: Angola, R. Longa, Chijija, *Baum* 620 (B†, holo., BM,
       COI, K!, iso.)

© The Board of Trustees of the Royal Botanic Gardens, Kew, 2005

# INDEX TO OCHNACEAE

## New names validated in this part

**Gomphia densiflora** (*De Wild. & Dur.*) *Verdc.* **comb. et stat. nov.**
**Gomphia likimiensis** (*De Wild.*) *Verdc.* **comb. nov.Ochna apetala** *Verdc.* **sp. nov.**
**Gomphia lunzuensis** (*N. Robson*) *Verdc.* **comb. et stat. nov.**
**Gomphia lutambensis** (*Sleumer*) *Verdc.* **comb. et stat. nov.**
**Gomphia mildbraedii** (*Gilg*) *Verdc.* **comb. et stat. nov.**
**Gomphia sacleuxii** (*Tiegh.*) *Verdc.* **comb. et stat. nov.**
**Gomphia scheffleri** (*Gilg*) *Verdc.* **comb. et stat. nov.**
  subsp. **schusteri** (*Engl.*) *Verdc.* **comb. et stat. nov.**
  subsp. **taitensis** *Verdc.* **subsp. nov.**
**Ochna apetala** *Verdc.* **sp. nov.**
**Ochna kirkii** *Oliv.* subsp. **multisetosa** *Verdc.* **subsp. nov.**
**Ochna leucophloeos** A. *Rich.* subsp. **ugandensis** *Verdc.* **subsp. nov.**
**Ochna polyarthra** *Verdc.* **sp. nov.**

© The Board of Trustees of the Royal Botanic Gardens, Kew, 2005

PLANTS PEOPLE
POSSIBILITIES

© The Board of Trustees of the Royal Botanic Gardens, Kew 2005

All rights reserved. No part of this publication may be reproduced, stored in a retrieval system, or transmitted, in any form, or by any means, electronic, mechanical, photocopying, recording or otherwise, without written permission of the publisher unless in accordance with the provisions of the Copyright Designs and Patents Act 1988.

First published in 2005 by
Royal Botanic Gardens, Kew
Richmond, Surrey, TW9 3AB, UK
www.kew.org

ISBN 1 84246 108 7

Design by Media Resources, typesetting and page layout by Margaret Newman,
Information Services Department,
Royal Botanic Gardens, Kew.

For information or to purchase all Kew titles please visit
www.kewbooks.com or email publishing@kew.org

# LIST OF ABBREVIATIONS

**A.V.P.** = O. Hedberg, Afroalpine Vascular Plants; **B.J.B.B.** = Bulletin du Jardin Botanique de l'Etat, Bruxelles; Bulletin du Jardin Botanique Nationale de Belgique; **B.S.B.B.** = Bulletin de la Société Royale de Botanique de Belgique; **C.F.A.** = Conspectus Florae Angolensis; **E.J.** = A. Engler, Botanische Jahrbücher für Systematik, Pflanzengeschichte und Pflanzengeographie; **E.M.** = A. Engler, Monographieen Afrikanischer Pflanzen-Familien und Gattungen; **E.P.** = A. Engler, Das Pflanzenreich; **E.P.A.** = G. Cufodontis, Enumeratio Plantarum Aethiopiae Spermatophyta; in B.J.B.B. 23, Suppl. (1953) et seq.; **E. & P. Pf.** = A. Engler & K. Prantl, Die Natürlichen Pflanzenfamilien; **F.A.C.** = Flore d'Afrique Centrale (*formerly* F.C.B.); **F.C.B.** = Flore du Congo Belge et du Ruanda-Urundi; Flore du Congo, du Rwanda et du Burundi; **F.E.E.** = Flora of Ethiopia & Eritrea; **F.D.-O.A.** = A. Peter, Flora von Deutsch-Ostafrika; **F.F.N.R.** = F. White, Forest Flora of Northern Rhodesia; **F.P.N.A.** = W. Robyns, Flore des Spermatophytes du Parc National Albert; **F.P.S.** = F.W. Andrews, Flowering Plants of the Anglo-Egyptian Sudan *or* Flowering Plants of the Sudan; **F.P.U.** = E. Lind & A. Tallantire, Some Common Flowering Plants of Uganda; **F.R.** = F. Fedde, Repertorium Speciorum Novarum Regni Vegetabilis; **F.S.A.** = Flora of Southern Africa; **F.T.A.** = Flora of Tropical Africa; **F.W.T.A.** = Flora of West Tropical Africa; **F.Z.** = Flora Zambesiaca; **G.F.P.** = J. Hutchinson, The Genera of Flowering Plants; **G.P.** = G. Bentham & J.D. Hooker, Genera Plantarum; **G.T.** = D.M. Napper, Grasses of Tanganyika; **I.G.U.** = K.W. Harker & D.M. Napper, An Illustrated Guide to the Grasses of Uganda; **I.T.U.** = W.J. Eggeling, Indigenous Trees of the Uganda Protectorate; **J.B.** = Journal of Botany; **J.L.S.** = Journal of the Linnean Society of London, Botany; **K.B.** = Kew Bulletin, *or* Bulletin of Miscellaneous Information, Kew; **K.T.S.** = I. Dale & P.J. Greenway, Kenya Trees and Shrubs; **K.T.S.L.** = H.J. Beentje, Kenya Trees, Shrubs and Lianas; **L.T.A.** = E.G. Baker, Leguminosae of Tropical Africa; **N.B.G.B.** = Notizblatt des Botanischen Gartens und Museums zu Berlin-Dahlem; **P.O.A.** = A. Engler, Die Pflanzenwelt Ost-Afrikas und der Nachbargebiete; **R.K.G.** = A.V. Bogdan, A Revised List of Kenya Grasses; **T.S.K.** = E. Battiscombe, Trees and Shrubs of Kenya Colony; **T.T.C.L.** = J.P.M. Brenan, Check-lists of the Forest Trees and Shrubs of the British Empire no. 5, part II, Tanganyika Territory; **U.K.W.F.** = A.D.Q. Agnew (or for ed. 2, A.D.Q. Agnew & S. Agnew), Upland Kenya Wild Flowers; **U.O.P.Z.** = R.O. Williams, Useful and Ornamental Plants in Zanzibar and Pemba; **V.E.** = A. Engler & O. Drude, Die Vegetation der Erde, IX, Pflanzenwelt Afrikas; **W.F.K.** = A.J. Jex-Blake, Some Wild Flowers of Kenya; **Z.A.E.** = Wissenschaftliche Ergebnisse der Deutschen Zentral-Afrika-Expedition 1907–1908, 2 (Botanik).

# FAMILIES OF VASCULAR PLANTS REPRESENTED IN THE FLORA OF TROPICAL EAST AFRICA

The family system used in the Flora has diverged in some respects from that now in use at Kew and the herbaria in East Africa. The accepted family name of a synonym or alternative is indicated by the word "see". Included family names are referred to the one used in the Flora by "in" if in accordance with the current system, and "as" if not. Where two families are included in one fascicle the subsidiary family is referred to the main family by "with".

## PUBLISHED PARTS

Foreword and preface
*Glossary
Index of Collecting Localities

Acanthaceae
    Part 1
*Actiniopteridaceae
*Adiantaceae
Aizoaceae
Alangiaceae
Alismataceae
*Alliaceae
*Aloaceae
*Amaranthaceae
*Amaryllidaceae
*Anacardiaceae
*Ancistrocladaceae
Anisophyllaceae — as Rhizophoraceae
Annonaceae
*Anthericaceae
Apiaceae — see Umbelliferae
Apocynaceae
    *Part 1
*Aponogetonaceae
Aquifoliaceae
*Araceae
Araliaceae
Arecaceae — see Palmae
*Aristolochiaceae
Asparagaceae
*Asphodelaceae
Aspleniaceae
Asteraceae — see Compositae
Avicenniaceae — as Verbenaceae
*Azollaceae

*Balanitaceae
*Balanophoraceae

*Balsaminaceae
Basellaceae
Begoniaceae
Berberidaceae
Bignoniaceae
Bischofiaceae — in Euphorbiaceae
Bixaceae
Blechnaceae
*Bombacaceae
*Boraginaceae
Brassicaceae — see Cruciferae
Brexiaceae
Buddlejaceae — as Loganiaceae
*Burmanniaceae
*Burseraceae
Butomaceae
Buxaceae

Cabombaceae
Cactaceae
Caesalpiniaceae — in Leguminosae
*Callitrichaceae
Campanulaceae
Canellaceae
Cannabaceae
Cannaceae — with Musaceae
Capparaceae
Caprifoliaceae
Caricaceae
Caryophyllaceae
*Casuarinaceae
Cecropiaceae — with Moraceae
*Celastraceae
*Ceratophyllaceae
Chenopodiaceae
Chrysobalanaceae — as Rosaceae
Clusiaceae — see Guttiferae
Cobaeaceae — with Bignoniaceae
Cochlospermaceae

Papaveraceae
Papilionaceae — in Leguminosae
*Parkeriaceae
Passifloraceae
Pedaliaceae
Periplocaceae — see Apocynaceae (Part 2)
Phytolaccaceae
*Piperaceae
Pittosporaceae
Plantaginaceae
Plumbaginaceae
Poaceae — see Gramineae
Podocarpaceae
Podostemaceae
Polemoniaceae — see Cobaeaceae
Polygalaceae
Polygonaceae
*Polypodiaceae
Pontederiaceae
*Portulacaceae
Potamogetonaceae
Primulaceae
*Proteaceae
*Psilotaceae
*Ptaeroxylaceae
*Pteridaceae

*Rafflesiaceae
Ranunculaceae
Resedaceae
Restionaceae
Rhamnaceae
Rhizophoraceae
Rosaceae
Rubiaceae
   Part 1
   *Part 2
   *Part 3
*Ruppiaceae
*Rutaceae

*Salicaceae
Salvadoraceae
*Salviniaceae
Santalaceae
*Sapindaceae
Sapotaceae
*Schizaeaceae
Scrophulariaceae

Scytopetalaceae
Selaginellaceae
Selaginaceae — in Scrophulariaceae
*Simaroubaceae
*Smilacaceae
Sonneratiaceae
Sphenocleaceae
Strychnaceae — in Loganiaceae
*Surianaceae
Sterculiaceae

Taccaceae
Tamaricaceae
Tecophilaeaceae
Ternstroemiaceae — in Theaceae
Tetragoniaceae — in Aizoaceae
Theaceae
Thelypteridaceae
Thismiaceae — in Burmanniaceae
Thymelaeaceae
*Tiliaceae
Trapaceae
Tribulaceae — in Zygophyllaceae
*Triuridaceae
Turneraceae
Typhaceae

Uapacaceae — in Euphorbiaceae
Ulmaceae
*Umbelliferae
*Urticaceae

Vacciniaceae — in Ericaceae
Valerianaceae
Velloziaceae
*Verbenaceae
*Violaceae
*Viscaceae
*Vitaceae
*Vittariaceae

*Woodsiaceae

*Xyridaceae

*Zannichelliaceae
*Zingiberaceae
*Zosteraceae
*Zygophyllaceae

-----------------------------------------------------------------------------------------------------------

## FORTHCOMING PARTS

Acanthaceae
   Part 2
Apocynaceae
   Part 2

Asclepiadaceae — see Apocynaceae
Commelinaceae
Cyperaceae
Solanaceae

Editorial adviser, National Museums of Kenya: Quentin Luke
Adviser on Linnaean types: C. Jarvis

Parts of this Flora, unless otherwise indicated, are obtainable from:
Royal Botanic Gardens, Kew, Richmond, Surrey TW9 3AB, England. www.kew.org or www.kewbooks.com

**\* only available through CRC Press at:**
UK and Rest of World (except North and South America):
CRS Press/ITPS,
Cheriton House, North Way, Andover, Hants SP10 5BE.
e: uk.tandf@thomsonpublishingservices. co.uk

North and South America:
CRC Press,
2000NW Corporate Blvd, Boco Raton, FL 33431-9868,
USA.
e: orders@crcpress.com

Information on current prices can be found at www.kewbooks.com or www.tandf.co.uk/books/